Teach Me To Read

The Early History of Lexia Learning Systems,
the Mission, and the Boy Who Inspired It All

by

Robert A. Lemire

Founder / Father

Dedication

This book is dedicated to all past, present, and future Lexia employees who have carried, are carrying, and will carry the Lexia mission forward.

This book is also dedicated to my son, Bo, who, although educationally ignored as a young dyslexic boy, completely overcame his reading disability and is now doing his part to carry Lexia's mission forward.

Library of Congress Control Number: 2015931402

Contents

Lexia President Nick Gaehde receives a preview of the *Teach Me To Read* manuscript from Bob Lemire

"Lexia was leading a 'revolution,' and people wanted to be part of it."

–ALICE GARSIDE

Foreword

It began as the effort to meet the needs of one child, but became a global mission to help millions of children acquire the reading skills needed to succeed in school and life. Bob Lemire believed that technology could be used to help struggling readers and children with language-based learning disabilities by providing intensive, rigorous instruction through research-proven methodologies. With tenacity and courage, Bob Lemire embarked on that mission, and for 30 years, he has been driven by the enormous responsibility to share this program with all who need it.

When he began Lexia, Bob didn't have a background in education, nor did he have a background in technology. In fact, he didn't begin this journey until he was in his 50s. However, Bob had an insatiable curiosity and surrounded himself with people who had the necessary expertise—and then ignited in them a passion and a commitment to helping children learn to read.

The clarity of Bob's mission drew people in. He was single-minded in solving the problem. Simply stated: Lexia helps students learn to read, enabling them to develop the skills that are absolutely critical to their success in life. Beyond the simplicity of this message, Bob also had the remarkable ability to help others understand the mission, internalize it, and make it their own.

I joined Lexia in 2005, after a lengthy conversation with Bob. The Lexia story resonated with me in a profound way because I struggled with dyslexia as a child, as did my children. I knew firsthand the power of Bob's mission, and the opportunity to join Lexia was a blending of my personal and professional ambitions.

The Lexia story is one of shared commitment and ownership of Bob's mission. He is one of those rare people with the ability to entrust their vision to others. As the company grew and evolved, a new wave of professionals joined because they saw Lexia as a career opportunity that also provided a level of personal fulfillment. We knew the incredible impact the company would have if we did it right—and that was exhilarating. As Bob's colleague Alice Garside once described it, Lexia was leading a "revolution," and people wanted to be part of it.

You won't meet anyone as tough and as kind as Bob. He simply refused to fail... he cared too much about children to let that happen. He has been tireless and relentless, measuring his success each day by the number of children he has helped. In the early days, reaching 100 students was an accomplishment. Today, Lexia is helping nearly 1.5 million students. However, that still leaves millions of students whom we have the responsibility to reach. Our work is far from complete.

I've always felt an overwhelming responsibility to help accomplish Bob's mission. It's not like some missions—where you don't know how to get to the finish line. The science of reading is well understood. We know what we need to do and, as Bob would say, "failure is not an option." Our mission is to continue bringing Lexia's proven pedagogy to more schools, more students, and more communities until we reach every student needing our help.

Bob would often compare Lexia's mission to the discovery and distribution of Penicillin. We just need to get it into the hands of those people who desperately need it. This is a life-changing program and we have the responsibility to share it with everyone who needs it—as if it were a life-saving medicine.

Teach Me To Read isn't so much of a retrospective of the first 30 years of Lexia. This book is a roadmap—one that illustrates not only where we've been as a company, but also where we need to go in the future.

As you read this book, I hope you become as inspired as those of us who had the life-changing opportunity to work with Bob, and that you make this story, this mission and this responsibility your own.

Nick Gaehde
June 2014

Foreword

Bob Lemire working on *Teach Me To Read*

Bob Lemire, Founder of Lexia Learning Systems

Preface

In 2012, when Lexia president and CEO Nick Gaehde asked me to write about the mission and early history of Lexia Learning Systems, I accepted his challenge as my final assignment. I am 80 years old now, and Lexia's story needs to be documented and shared before I retire.

I helped found Lexia in 1984 shortly after our son, Bo, was diagnosed with developmental dyslexia. Bo's diagnosis and remediation were the essential reasons Lexia was founded. Our intimate experience with the pain and frustration of seeing a child with reading difficulties move through the educational system gave us the strength, courage, and motivation to create Lexia and its reading products.

This book thus contains two intertwining stories. One is about a dyslexic boy who had trouble learning to read. He is our son but could have been any child experiencing reading difficulties. My hopes and dreams are that Lexia's products will keep other children from having the same painful experiences.

The other story is about the actual development of Lexia Learning Systems. It is about observing a problem, understanding it, and developing a solution. Lexia's founding mission was to end reading failure for dyslexic children and, later, for all beginning and struggling readers in the English-speaking world. The story is also about the difficulty of finding investors who believed in Lexia's mission, as well as the struggle to convince the educational community that Lexia's products could help them succeed in teaching reading.

Had I known that writing the history of Lexia Learning Systems would eventually be my task, I would have made notes and saved documents along the way. Not having done so, I asked my spouse, Virginia, for help and together we approached our son to learn about his early student experiences and memories. Bo's responses were sometimes hard for us to hear:

"I was diagnosed with developmental dyslexia in fourth grade. My disability changed my life forever. It also changed my father's life forever. Most of my early childhood memories are buried and difficult to bring back, based on the suffering I experienced as a poor student. It affected every aspect of my life growing up."

–Bo Lemire

Virginia and I also interviewed many others involved in Lexia's history, including about 30 current and former employees, a member of Lexia's Board of Directors, and independent resellers. Their personal statements and stories further humanize Lexia's history.

It is my intent that readers of this book will come to understand some of the causes of reading failure as well as an effective and affordable means of overcoming or even preventing it. You will see that reading failure need not be a matter of destiny for any child.

I hope that I am able to convey all the passion and dedication that were poured into the creation of Lexia Learning Systems since its beginning.

"I have no memory of learning to read;
it just happened."

—Bob Lemire

Chapter 1
About the Founder

Looking back, I wonder what skills I possessed that would make it possible for me to help start any company, let alone one focusing on a subject about which I knew nothing. Perhaps I just made the first move and others followed. I certainly did not know how difficult it would be.

I was not a teacher, not an educator, and not a technical person. I had never managed employees, had never personally raised venture capital, and had no interest in "going out on a limb" to start anything new. I already had more than one rewarding career and was not looking for another.

I have no memory of learning to read; it just happened. I knew nothing about the acquisition of reading, although I was a good reader. My family, neighborhood, and grammar school were bilingual so I learned to speak, read, and write in French and English. My grandparents, though fully literate in English, preferred to speak French. My parents spoke French and English with equal facility.

I attended a Franco-American Catholic elementary school in Lowell, MA within walking distance of our house. In the lower grades classes were taught by nuns and in upper grades by religious brothers; all were taught in French and English. The teachers drove us very hard. One day in third grade I forgot how to spell the French word for the number "10" and was sent to the first grade to ask the class how to spell the word "dix." I have never forgotten that experience. Perhaps this humiliation is what makes me feel so much compassion for the reading-challenged.

School was very easy for me, though I certainly did study and work hard. When it came time for high school, I wanted to attend the public Lowell High School, where my father taught mathematics. But there was also a Catholic high school in Lowell where all Catholic students were expected to attend. In order to go to the public high school, my father was required to seek an exemption from Cardinal Cushing. Permission was granted and I attended Lowell High School, where I studied more French, Latin, and, in my father's classes, math. I recall often waving my hand and saying "Hey Dad! Dad, I know the answer!"— to the annoyance of my classmates. I also found sports easy, especially football. I became a successful player and was an "All-State" player my senior year.

Yale accepted me as a full-scholarship student and I graduated with a BA degree in economics. I also became an Ensign in the US Navy and went on to gunnery school to prepare for commanding a gun turret on the heavy cruiser, the *USS Baltimore*. After two years in the Navy, I returned to New England to attend Harvard Business School on the GI Bill, graduating with an MBA degree in 1958. After writing case studies for a Boston consulting company, I was hired by Paine, Webber, Jackson, and Curtis, a well-known Boston brokerage and financial firm, to work in their corporate underwriting department. Looking back, it is fortunate that I spent a few years at Paine, Webber because I learned the legal requirements to fund and launch a new enterprise. This eventually proved to be very helpful in the founding of Lexia.

Virginia and I eloped in 1960 and lived in Cambridge for a year before purchasing a house in Lincoln, MA. We knew little about the town except the good things, including its well-regarded school system, that we had heard from our coastal-Maine honeymoon host, Richard Thorpe. Dick was himself a Harvard Business School graduate and a long-time Lincoln resident. He and his wife introduced us to many of their friends from Harvard, MIT, and other universities. For several years I commuted daily on the train to Boston, which provided the opportunity to get to know other Lincolnites. Lawyers, doctors, and businessmen converged daily on the Lincoln train platform, and I became well acquainted with many of them by name and profession. I soon realized that many Lincolnites were leaders in their fields.

Our daughter, Elise, was born in 1964 and our son, Robert B., nicknamed "Bo," was born in 1968.

On my train commutes to Boston, I became close friends with several town-government volunteers. I was soon appointed to the Lincoln Conservation Commission, joining the ranks of numerous volunteers who spent their spare time governing the town. During my 15 years as chairman of the Lincoln Conservation Commission, the town put some 1400 acres into permanent conservation. My Lincoln land-use experience led to consulting requests from the Nature Conservancy, the Conservation Foundation, and other national organizations. Various conservation groups from across the country hired me for advice and on-site consultation for land planning and conflict resolution. Languages of the law, finance, and land were key ingredients in my success at providing landowners with fair after-tax market value while preserving the land-use interests of the larger community. In 1979, Houghton Mifflin published my book, *Creative Land Development: Bridge to the Future,* to help other communities understand the tools and techniques available to conserve their town's undeveloped land while building what needed to be built to serve community needs.

In the mid-1970s, I left Paine, Webber to start my own one-man private investment-advisory service. Lemire & Company initially hired the services of a small company to run the monthly client-account evaluations on a time-shared mainframe computer. In 1977, I moved my investment-management business from Boston to Lincoln and purchased an Apple II computer with appropriate software to run the periodic portfolio evaluations myself. Virginia operated the computer and did some simple programming. We became quite dependent on our Apple and even purchased another Apple II in case we had a hardware failure.

During the day I worked in my investment-management business, and pursued my conservation interests nights and weekends. I also taught evening land-use classes at the Conway School of Landscape Design (Conway, MA) and in the landscape-architecture departments at Harvard University and Rhode Island School of Design.

By 1981, I had these two businesses and teaching going quite well. Those were rewarding years for me. Life was good and I had no reason to change anything.

Suddenly a change occurred that was to affect the rest of my life.

*"When the teacher called, it was not
good news."*

–B<small>OB</small> L<small>EMIRE</small>

Chapter 2
A Call from Bo's Teacher

The Lexia story began in 1978, when Bo was in the fourth grade. Out of the blue, his teacher telephoned and told us that Bo needed to be seen by a psychiatrist. She was sure he was having a nervous breakdown.

Bo had had two years of private preschool before starting kindergarten in the local public schools. He was a bright boy with successful report cards and teacher evaluations. There were hints of reading problems, but the school was not providing him with any extra help. If the school was not concerned, then neither were we. Only in the following years did we learn how wrong it had been to ignore those early signs.

The year before, in third grade, Bo had started to have some behavior problems, and his teacher telephoned us several times over the course of the year to discuss his progress. Neither Bo's third-grade report card nor the telephone calls from his teacher said anything about reading difficulties. This teacher praised his creativity, enthusiasm, and participation, and reported that Bo was very popular and was developing strong leadership qualities. She also reported that Bo did not recognize these talents in himself and refrained from using his developing skills.

HARTWELL-SMITH SCHOOL   PAGE 1

NAME Bo Lemire _____   DATE June 1977

ALL areas are considered in each child's development. The teacher comments only on those areas which need clarification.

PERSONAL AND SOCIAL DEVELOPMENT	Seldom	At Times	Almost Always	COMMENTS
Is considerate of others			✓	
Is responsible			✓	
Is cooperate			✓	
Is attentive			✓	
Interacts well with peers			✓	
Interacts well with adults			✓	
Accepts criticism			✓	
Shows interest in learning			✓	

Comments: Bo is one of the most sought after children in the class. At recess everyone wants to be on his team – in class everyone wants to befriend him. He on the other hand prefers working with one or two others – close friends or by himself. He is a most independent child. It has

Conference Comments:

truly been exciting to see him grow more confident and aware of his capabilities. It has effected positively every aspect of his work.

Teacher:

HARTWELL-SMITH SCHOOL   Page 2

NAME Bo Lemire _____   DATE June, 1977

LANGUAGE ARTS
READING
Word Attack improved
Oral Reading smoother
Comprehension good
Reading for Enjoyment
Work Habits

COMPREHENSION LEVEL: Below Grade / (At Grade) / Above Grade

WRITING SKILLS
Written expression
Mechanics of writing Cursive well
Handwriting developed –
✱ Spelling much improved
Capitalization yet continues to
Punctuation need much
Sentence Structure reinforcement
Paragraphing reminding &
✱ Proofreading can and much (teaching
WORK HABITS to be encouraged to
ORAL LANGUAGE do so
LISTENING ABILITY

COMMENTS

Bo has made good progress over the course of the year. His word attack skills are substantially stronger, his oral reading smoother. He reads willingly assigned and non assigned material.

Until recently written work was something he avoided as much as possible – I have been excited to see this attitude change. He now writes with but little hesitancy and is becoming increasingly more adept at expressing in writing what he wishes to say.

Teacher

Third grade report card

Looking back, I think we knew little of how Bo was feeling and performing. We did not support him emotionally in the way he may have needed. We thought principally of his education, believing it was what would support him throughout his life. Hindsight can be very painful.

"School was fun and exciting through the third grade. In the fourth grade, I recall having trouble with the work and being told I was not trying hard enough. If you hear that enough, you start believing it. It seemed to me that I was trying just as hard as the other kids, so I must have been stupid. If I was not good at doing the assigned school work, I would find something I was good at, such as disrupting the class, fighting with other kids, and disrespecting teachers. I sure was good at that."

—Bo Lemire

In fourth grade, Bo was in a combined fourth- and fifth-grade class. The double-sized classroom had nearly 50 students, two teachers, and one or two aides. This turned out to be a difficult year. When Bo's fourth-grade teacher called us with her alarming request to take him to a psychiatrist, she had no idea what might be wrong. Virginia was particularly stunned by the teacher's statement.

"I had my own experiences with 'nervous breakdown.' The teacher's call terrified me. After the birth of both of our children, I had post-partum depressions and knew how completely desperate someone can feel during a mental breakdown. If Bo felt anything like I had felt, he was suffering terribly. He didn't seem to be suffering, but on the other hand he was not the happy boy we knew in prior years. What was the problem? We couldn't ask him because he didn't know. I wept for him."

—Virginia Lemire

What course of action should we take? We had a good relationship with Bo's pediatrician but he was not a psychiatrist. Then I remembered that one of the men riding the train to Boston was a doctor at Massachusetts General Hospital who frequently talked about children. The next morning I found him on the train platform and asked for advice. Dr. Edwin Cole said that he was a psychiatrist and suggested that we schedule an appointment with him in his Boston office.

"Little did I know that I had approached one of the country's most respected figures in the area of reading disabilities and remediation."

—BOB LEMIRE

Chapter 3
Meeting with Dr. Cole

At our appointment, Bo went with a staff member to the language-clinic laboratory for testing while Dr. Cole interviewed Virginia and me. He asked several questions about Bo and his use of language. For example, was Bo left-handed? We responded that he was not. (Later we were told that there is a higher incidence of dyslexia among left-handed people.[1]) Did Bo use words or phrases that were not quite right? Yes, we replied, he said things like "whobody's at the door" and "basghetti." At the end of the testing, we and Bo were told that he had developmental dyslexia and that it was characterized by difficulties in accurate and/or fluent word recognition as well as by poor spelling and decoding abilities in otherwise healthy persons. Dr. Cole told Bo that "all of this is not your fault." He could be treated with the help of tutors and he would learn to read. We knew nothing about dyslexia, but we were certainly eager to learn. I suddenly had a new mission in life—to teach our son to read. But I didn't know how to do that, and had to rely on Dr. Cole.

We delivered Dr. Cole's diagnostic reports to the school office and to the fourth-grade teacher who had requested his psychiatric evaluation. At Dr. Cole's direction we engaged a private tutor for the few remaining weeks of

---

1. Norman Geschwind, M.D. "Samuel Torrey Orton" Annals of Dyslexia, Vol. 32, 1982:16

the school year. For the summer break, Dr. Cole made arrangements for Bo to be tutored by Angela Wilkins, a teacher at the Carroll School in Lincoln. Bo blossomed under this one-on-one instruction. Not only was he eagerly reading his assignments, but he seemed to be a much happier boy. According to Ms. Wilkins, he had a long way to go before he would read at grade level[2] but he was taking the first essential steps toward remediation of his dyslexia.

What do parents do who have no "train-friend" who specializes in reading and children? It was a complete accident that we had found *the* person to evaluate and guide Bo's future education. Lexia was founded because such an expert was not available to every parent.

The history of Lexia would not be complete without including Dr. Cole's history as a reading expert. Who was this kind and gentle man? Through friends, I contacted Dr. Cole's children, Abigail Cole Dawkins, Timothy Cole, and Jennifer Cole Gould, and asked them to tell me about their father. The following was taken from their responses.

Edwin Cole was born in Cohasset, MA on June 18, 1904, the third child of a Unitarian minister. When Edwin was 15 years old, his father died suddenly, leaving the family with virtually nothing. Although the church was helping them, his mother needed to find a job. She soon secured a position as a house mother at the University of California at Berkeley.

Edwin completed high school in Berkeley and studied for two years at the University of California at Berkeley. He transferred to Harvard University, supported by what is now called the Children of Unitarian Universalist Ministers College Scholarship. Each year Edwin had to request tuition benefits from the fund for the next year, an exercise that was extremely uncomfortable for him. In his senior year at Harvard, while taking a mandatory anatomy course, Edwin chose medicine as his future career. His path would not be easy: he spent a fifth year taking all the science courses required to enter Harvard Medical School. During medical school, Edwin had a benefactor and worked at the First Unitarian Society in Boston.

After graduation, Dr. Cole undertook his internship and two residencies, one in psychiatry and the other in neurology. In 1934, he served several months in New York City as a research assistant with Dr. Samuel T.

---

2. A child is on grade level if he or she is performing in the same way as other children his or her age.

Orton, a neuropsychiatrist who had pioneered the study of learning disabilities in the 1920s. Dr. Orton, with educator and psychologist Anna Gillingham, formulated an orderly, systematic approach to the English language. It is likely that Anna Gillingham was working with Dr. Orton when Dr. Cole served as his assistant. Dr. Orton's name has come to be strongly associated with the Orton-Gillingham (O-G) method which remains the basis of the most prevalent form of remediation and tutoring for children with dyslexia or dyslexia-like symptoms, such as reading disabilities.

In 1934, Dr. Cole returned to Boston and opened his office at 311 Beacon Street, specializing in neuropsychiatry. Simultaneously, he founded the Language Clinic and the Cortical Function Lab, both at Massachusetts General Hospital. He utilized the Orton-Gillingham method in his private practice, while the language lab trained teachers and tutors in the O-G method. Before long, his method for teaching students with language-based learning disabilities, including dyslexia, was embedded in several Boston-area private schools, including Exeter Academy, Andover Academy, Dana Hall, and the Cambridge School of Weston.

Dr. Cole was to remain at the Language Clinic for 50 years. When he and his wife, Lucy, moved to the exurb of Lincoln, MA in the early 1940s, he commuted to Massachusetts General Hospital by train for many years.

In 1967, Dr. Cole helped found the Carroll School in West Newton, a school for children with language-based learning difficulties. The school quickly expanded and was moved to Lincoln. Dr. Cole also served as a consultant and/or trustee at several other schools. For families with dyslexic children who lived in areas where no special schools were available, he offered summer boarding-school opportunities at Exeter Academy (NH), the Cambridge School of Weston (MA), and St. George's School (RI). For many years, he traveled weekly to the St. George's program, to monitor the progress of his patients.

Upon Dr. Cole's retirement, in 1984, his patient records were given to the Carroll School.

As parents, Virginia and I had no idea what Bo needed in order to read at grade level and ultimately become a successful reader. What had happened in the lower grades, when he missed the acquisition of essential reading skills?

What were the essential skills he had missed? The answers to these questions would later become the framework for starting Lexia Learning Systems. But at the time, we knew little about what Bo was learning with his tutors, and we would remain uninformed for some time to come.

Bo was a quiet boy and had only a small handful of friends with whom he enjoyed playing. When we had a surprise party for him on his 11th birthday, inviting his close friends (both boys and girls), Bo seemed very uncomfortable. Knowing what we do now, we can understand his reaction; but we never asked him about it, either at the time or since.

Dr. Edwin Cole

"Bo should be doing better."

—A NOT-INFREQUENT TEACHER COMMENT

Chapter 4
Bo in Fifth and Sixth Grades

Early in Bo's fifth-grade year, a group of teachers associated with his education gathered to review his situation. This group included Mary Mendler, Bo's private O-G tutor whom Dr. Cole had recommended, Bo's classroom teachers, and the school's special-education teacher.

"I attended this review and was struck by the description of Bo's abilities and disabilities. I recall not wanting to participate in the discussion but, instead, wished I could crawl under the table and just listen. Their description of Bo could have been a description of my own abilities and disabilities. I was shaken by this information and wondered what it meant. Did I have the same disability as Bo?

"A year or two later I had to fulfill the language requirement for my master's degree by showing reading proficiency in a foreign language. I had a hard enough time reading English, let alone some other language. I consulted with Dr. Cole, who said he was sure I was dyslexic, just like Bo. He wrote a letter to Boston University and I was exempted from the requirement. I now realize that my father was also dyslexic."

— VIRGINIA LEMIRE

The outcome of the teachers' review was that there was nothing the school planned to do for Bo except provide extra spelling help in the "corner room."

> "What an unfortunate name for the special-education room. Isn't it just like saying to students who need extra help, 'Go sit in the corner'?"
>
> −Bo Lemire

We learned much later that Bo had received his reading instruction in some form of the "look and say" method, now called "whole language." Simply stated, the whole-language approach involves exposing students to reading until they just "get it." Bo became a victim of the whole-language method. There is more about whole language later in this book.

The aides in the corner room were not reading specialists. The O-G reading instruction Bo received from his tutors was not being reinforced in his classroom or in the corner room. Spelling help with the corner room aides entailed the use of word lists—and for the students who did not "get it", (according to whole-language methods) these lists had to be memorized. This was the same type of instruction Bo was getting in the regular classroom; how was it supposed to help him? For most of his sixth-grade year, Bo spent entire class periods in the corner room, with little to show for it except a strong affirmation that he had a learning disability.

Halfway through Bo's fifth-grade year, the First Parish Church in Lincoln produced the children's Christmas opera *Amahl and the Night Visitors*. Bo sang the part of Amahl, the young crippled beggar whose infirmity was miraculously cured when he sent his crutch with the Three Kings as a gift for the Christ child. This production was very emotional for Virginia and me, as well as for Bo's O-G tutor, Mary Mendler, and his Carroll School tutor, Angela Wilkins, who attended with us. Bo recently told us how much he enjoyed being in that production and how many astounding compliments he received for his voice and acting abilities. He also said it had been easy to memorize all the material because learning music was completely different from learning written language.

In the first reporting period of Bo's sixth-grade, the teacher was well aware of his diagnosis when she gave him an F in vocabulary and an F in grammar. Did she think this would motivate him? His language-arts assessment stated that he "should be doing better." These were the same words we would hear from Bo's teachers all through middle school and high school, and he came to hate them. He did well in verbal communication and oral classroom work, but less well on written work. In other words, he was very intelligent but performed poorly on written tests.

Sixth grade report card

Most teachers are able to recognize reading difficulties by the time a student has passed fourth or fifth grade. But understanding and remediating reading difficulties were not part of their training. One member of Lexia's Board of Directors told me how he had learned to teach reading successfully.

"Prior to 1977, I taught in a Kentucky elementary school. I thought that non-readers in my classes had just not met a teacher who was demanding enough to get them reading. I could not accept that these students could not learn to read by the same instruction under which other children in my classroom learned to read. If these students needed to have a demanding teacher, I would be that teacher. So for two years I was the meanest, most demanding teacher I could be—but they did not learn to read. All I did was make them cry. It was horrible for them and for me. Thank goodness the Orton-Gillingham (O-G) method for teaching reading accidentally came to my attention. I relentlessly pursued the study of this method, even moving to Boston for O-G training, until I was the teacher I wanted to be and my students learned to read successfully."

— EARL OREMUS, HEADMASTER, MARBURN ACADEMY

The Brook School in Lincoln, MA

"I hated school."

—Bo Lemire

Chapter 5
Bo in Seventh Grade

In seventh grade Bo was so unhappy in school that we asked Dr. Cole if he might be better off at a private day school. He advised us to look into it. We agreed to visit two schools in Cambridge that might be appropriate and arranged appointments. At one school, Bo spent an afternoon in a seventh-grade classroom:

> "I took Bo for these visits. After attending classes for the afternoon at one school, I asked him how it went and what he thought. After some silence, he described one class where he thought the teacher was not very good because several boys tore all the buttons off another boy's shirt, one at a time, during the class hour. I saw clearly that he did not want to be in a school where some students freely picked on other students."
>
> —Virginia Lemire

At the end of seventh grade, Dr. Cole talked to Bo again and reviewed his progress. He also spoke with Bo's tutor and learned that Bo still had some catching up to do to be a successful reader. Even with a tutor three hours a week, his reading ability was not at a level consistent with the norms for his age and grade. Dr. Cole recommended that Bo attend a school that specialized in learning disabilities, reasoning that Bo would benefit much more from remediation in a boarding-school environment tailored particularly to his needs.

Bo had certainly made progress with his tutors, but by the time a student is in the seventh grade, it is critical that he or she be able to read to learn. Reading problems are much more difficult to remediate without the right kind of help, and many public schools were (and still are) unaware of or unable/unwilling to provide that help. Junior-high curriculum had moved on from learning-to-read to reading-to-learn. Virtually all the help that Bo had received since fourth grade was in the form of private tutoring arranged by Dr. Cole and paid for by us. It had done a lot for Bo, but there was a lot more to be done. According to Dr. Cole, a year or two in an appropriate school would permit Bo to catch up on essential reading skills while surrounded with students in similar or identical situations. Virginia and I were, nonetheless, very upset to think about sending Bo away.

These are the emotional and financial costs of delayed remediation, and it is these experiences that motivated us as we started Lexia. We wanted to save other students and parents from having to go through what Bo, Virginia, and I had experienced.

> "Having difficulty with reading had an impact on every other aspect of my life. The older I got before getting real help just made the impact worse. Explicit instruction in reading must be available in the early years, before other problems arise."
>
> – Bo Lemire

The Carroll School in Lincoln was ruled out because it did not admit older students and was not a boarding school, which we were persuaded would help Bo with the many subtler issues of long-term reading difficulties. We know now that older students who receive remediation in late elementary grades or after often develop other problems, including low self-esteem, that require attention. By the time a student is 11 or 12 years old and is not a grade-level reader, he or she very likely experiences his struggle as a reflection of something wrong in himself, that he may be "stupid" or "dumb," and that no matter how hard he tries, he cannot do school work as easily or as well as others.

"I had an invisible disability. What is it like to have this particular problem? I wanted to be like everyone else but couldn't. Everyone wants to fit in with their peers, to be part of a group, especially as a young kid. I didn't fit in with my peers in the classroom, which was obvious. I was unable to complete the schoolwork my peers were finding easy. It is difficult to be the stupid kid on the bus, or at a birthday party, or on the playground. The teachers treated me as if my disability was just a character defect easily corrected with stricter rules or longer time on task.

"Eventually it is easier to stop trying and become what you are perceived to be. It becomes easier to be lazy, unmotivated, and poorly behaved. It makes more sense to be the bad kid. If I can be good at something, I can be good at being a bad kid. Who do bad kids play with? Who do they hang out with in school? What do bad kids do in their free time? You probably know the answers to these questions."

-Bo Lemire

Dr. Cole recommended that Bo attend the Greenwood School, a private boarding school in Putney, VT. It was a relatively new school with a teacher/ student ratio approaching one to two. Dr. Cole had advised the school at its founding, in 1978, and served on its Board of Trustees. As parents, we were conflicted. Why did parents have to send their child away to boarding school to learn to read? We wanted the best for Bo, but sending him away was very difficult for us. We would lose our parenting opportunities, lose the chance to demonstrate the human and social values that parents teach their preteens and teens. Bo would have to learn these things without us. The thought of not having daily contact with our twelve year old son was devastating.

We wondered for the millionth time why we had not known about and addressed Bo's reading problem long before the fourth grade. Were we poor parents not to have recognized his issues much earlier? Might we have avoided this present monumental change in all our lives? Might we ever know how Bo was going to feel about boarding school?

The three of us visited the Greenwood School in the spring. The total enrollment was 28 boys of elementary and junior-high ages. Trusting Dr. Cole, we made an application and Bo was accepted.

"The decision to go to Greenwood was like admitting defeat. No matter how the adults framed it ("You'll love it! They'll teach you what the other teachers can't! You'll be around kids just like you!'), it was still a school for "different" kids."

—Bo Lemire

Beyond these personal and familial concerns, there was the cost. In 1981, annual tuition at Greenwood was over $28,000. In 2012 dollars, tuition at a comparable boarding school was about $65,000.

"At the time, we were paying tuition at Boston University for me to pursue a Master's degree in computer science which, we hoped, would help me find a good job to finance our children's college costs. Nevertheless, if Bo needed to go to this special school, we would do it somehow."

—Virginia Lemire

We considered asking our local public school to conduct an educational plan, called a "766 evaluation," which might result in the school district's paying some or all of the Greenwood tuition. Enacted in 1972, Chapter 766 of the Massachusetts General Laws guaranteed the right of all special-needs students (ages 3-22) to an educational program best suited to their needs. Under Ch. 766, special-education teams were to be convened and evaluations and annual reviews were to be conducted, all toward developing an ongoing, individual education plan that ensured an appropriate education. Local school systems were required to educate every student in their community and be responsible for funding appropriate educational costs.

Our school did not consider Bo a "special needs" student. Though he had received all his elementary education in the local school, and though his reading ability was therefore entirely the product of the school's teaching, the school denied that it had failed to make him a successful reader. His disability was still invisible to them.

As we spoke with parents who had taken the "766 path," we learned that the process was so slow that Bo might remain in the local school for one or two more years. We decided to proceed on our own.

In fairness, I don't think the Lincoln Public School's treatment of Bo was altogether unusual for the times. In the late 1970s, I believe, most public schools would have done much the same in a similar situation. Further, children can often camouflage their lack of reading skills, especially if they have an exceedingly good memory or if they become otherwise adept at masking their difficulties. Such "camouflaging" is reported to continue still, as high intelligence is cunningly used to cover up reading problems and superb memory is used simply to store words rather than decode them. This is why reading problems may not fully reveal themselves until third, fourth or even fifth grade—when a student's camouflaging techniques are exhausted.

Clearly, if schools make no attempt to test their students' decoding skills[3] in first, second, and third grades, reading problems may not be discovered early enough. And if left untouched until fourth grade or beyond, reading issues are very difficult to remediate. I know this firsthand.

We brought Bo to Greenwood in early September of 1981. Vermont was beautiful in the autumn, but dropping off our 12-year-old was very stressful for all three of us. The school had a number of activities on that day to ease our tension. Each "new boy" had an assigned "old boy" to help him during his orientation. As we left the school, Bo's assigned buddy took him to their next activity.

"I don't think Bob and I said a word all the way back to Lincoln."

-VIRGINIA LEMIRE

---

3. The ability to de-compose words into their sounds.

The dining hall at Greenwood School

"I figured I was not quite as dumb as ALL the other Greenwood kids, like I'd felt in public school."

–Bo Lemire

Chapter 6
Bo at Greenwood

We were completely separated from our son until Thanksgiving. Bo's September birthday came and went without our being able to celebrate with him. Students were not allowed on-campus or off-campus visits until Thanksgiving, as the school felt many parents indulged their disabled child too much. While we did not think we over-indulged Bo, it quickly became clear that some parents did: in the "Boston carpool" that we participated in at the start and end of vacations, we observed many attempts to smuggle huge bags of candy and treats into the Greenwood dorms, only to have them confiscated by the staff.

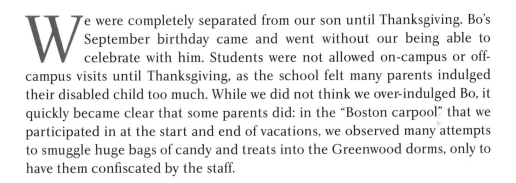

"I was already a quiet kid when I arrived, so I stuck with that acquired skill as I set out in my new school. I remember thinking I was at a school for kids who can't do the work at a regular school. I wondered how it would pan out."

–Bo Lemire

We eagerly visited the school when parents were invited. Bo said little to us about what he was doing.

"I had some athletic skills that many of the other kids didn't, so I got to feel good about myself in a few areas of my life."

—BO LEMIRE

Bo learned to cross-country and downhill ski while at Greenwood. The school could be flexible enough with its schedule to permit skiing on a mid-week day and have classes on Saturdays. Besides classes, the boys participated in housekeeping and maintenance chores at the school. At the end of the year, Bo's report card was very good. He was looking forward to the summer break.

"During that summer, Bo spent a few weeks in Illinois on my sister's family farm. Including Bo, there were five boys, ranging in age from 12 to 20 years. All of them participated in the large-scale farming operation. Bo learned to drive a huge diesel tractor, sometimes hauling a 40-foot-wide cultivator that, at the touch of a button, he could make fold up to eight feet before driving it through a gate between fields, and then unfold again. That was heady stuff for a 13-year-old! It helped him feel good about himself."

—VIRGINIA LEMIRE

Bo returned to Greenwood in the fall of 1982. He had had a good summer, but returning to the school for "different" students was still not where he wanted to be.

"Why did we have to send our son away from home to learn to read?"

Chapter 7
The Idea

In 1982, I was asked to serve as a trustee of the Greenwood School. Both Dr. and Mrs. Cole were trustees at the time, so we could drive together to Vermont for meetings. As part of my orientation, I visited various classes, one of which was a tutorial where I observed a teacher drilling a single student with flashcards. I was told that every student needed several of these one-on-one tutorials each week to master the skills needed to become a proficient reader. What were these skills and what were the flashcards? Little did I know that the experience of seeing this tutorial would change not only Bo's life and my own, but those of other struggling students as well.

I was full of questions for Dr. Cole on our return trip to Lincoln. I asked him to explain the tutorial and the flashcards. He replied that Dr. Samuel Orton and Anna Gillingham had deconstructed the English language into its essential components, which came to be known as "the code." I had been watching the code being taught to a Greenwood student using the Orton-Gillingham method. Dr. Cole explained that the English language is made up of 44 sounds and that there are 150 letters and letter combinations to make the 44 sounds. For example, the f sound in friend is also made by the ff in different, the ph in telephone, and the gh in enough. Single letters may also have more than one sound, like the o in hop and hope, or the g in gain and gin. I had never thought about any of it, even though I used this subconscious knowledge whenever I wrote or read.

The conversation with Dr. Cole led me to more investigation, which in turn led me to learn about how we acquire reading skills. I learned that there are five short vowel sounds, six long vowel sounds, 18 consonant sounds, seven digraphs[4], three r-controlled vowel sounds, and several diphthongs[5]. The English-language code was surely complicated, and not easy to learn, but the tutor I had observed was teaching the code one skill at a time: as each skill was learned, the next skill was presented. Over time, the student learned the entire code—44 sounds and the 150 different ways to represent them. I finally started to realize how English-language skills could be difficult for some students to master, and I became passionate about how people learn to read.

I learned that knowing how to "decode" provides the understanding of letter/ sound relationships and letter patterns in order to pronounce familiar words quickly and figure out other words that the reader has never seen before. The patterns of letters in a syllable are an extremely important part of O-G, as they inform the reader how to pronounce the vowel sound. When learning to read, some students figure out these relationships on their own, but most students do better with explicit instruction. Bo had not figured this out on his own, but his one-on-one tutorial was helping him make up for lost time.

I learned that reading is a recently acquired human skill. It is not part of our DNA, like walking and talking, for which our bodies are programmed in our genes. Although we are not born predisposed to read, our brains can develop the neurology needed to process new skills, including reading. Basic reading capability is achieved by receiving small increments of learning at the boundary of one's existing knowledge. Repetitions grow the neural networks in the brain that will automatically apply these skills and thereby enable our brains to process written words for the information they convey. Without these decoding skills, the student is forced to commit each word learned to his/her permanent memory. With these decoding skills, the student has only to memorize some 200 irregular, nondecodable words ("sight" words) to permanent memory, and then can simply apply the

---

4. A group of two successive letters whose phonetic value is a single sound, such as ai in rain or ay in day.
5. A sound made by combining two vowels, such as o plus i in boil.

code to attack new words. It's called "sounding out words." Once you can decode quickly and without mistakes, the requisite neural infrastructure is fully in place. Once the essential decoding skills and sight words have been learned, you can read to access the knowledge that the words represent.

I learned that reading is a complicated matter, but that once you can apply the code and the sight words, you can deal with each and every English word that you confront.

–You can say it.

–You can write it.

–You can think about it.

–You can learn from it.

–You can use it again and again.

–You can look it up in a dictionary.

I learned that there are several hundred thousand words in the English language, and that the human memory cannot possibly hold all of them in word form. Phonological knowledge, the knowledge of the sound structure of language, enables a person to quickly deconstruct written words into their sounds and know them for the meanings they convey. Once a student can automatically decode at reading speed, his or her brain is free to concentrate on comprehension.

Until quite recently, reading was not that important. It was not until the late 1940s or early 1950s that the rate of 18-year-olds graduating from US high schools surpassed 50%. Part of the reason was that most jobs did not require reading skills, and so most school children were required to learn only what was necessary for their future trades; their vocabulary words were simply memorized. That was not what we wanted for Bo. We told him that having an honest menial job was alright with us, but we wanted him to make himself open to other choices.

The one-on-one tutorials and flashcards were helping Bo build the neural networks that would make his word recognition automatic. Some skills required few iterations, others required many, but the practice would continue consistently until Bo stopped making mistakes. Once he mastered the code, he would be able to decode any new word other than the couple hundred irregular words that do not follow the code; he would memorize those nonconforming words for future sight recognition. Once he could automatically decode conforming words and sight-read nonconforming words, he would be a proficient reader.

The O-G flashcards in use at Greenwood set me to thinking more and more. The heavy one-on-one tutorial time was much of the reason the school was so expensive. We were lucky to have Bo there—but I thought that not many other parents could afford a school like Greenwood. So how could that tutorial be made available to more students like Bo? I thought about the small computer in my office that did repetitive calculations and printed out the results. No one wanted to do that by hand when a computer could do it more reliably, more quickly, and at less cost. There it was: I saw the one-on-one tutorials with flashcards as the same sort of expensive repetitive work.

Suddenly it struck me: could a computer program be created that could (i) offer its infinite patience to a student no matter how long it might take the student to learn the skill, (ii) keep a record of each response from the student and automatically repeat the same skill exercise until the student got it right, and (iii) advance the student to new and more difficult skills as each skill was learned? Was anyone using the computer's greatest attributes in the field of education?

I felt rather frightened to think of such a revolutionary idea. I was not a technical person. I would need help.

"Was there any alternative to special one-on-one tutoring for students like Bo to learn to read?"

—Bob Lemire

Chapter 8
Meeks Associates

Seeing our son acquire reading skills through one-on-one tutoring at Greenwood, and having learned so much myself from Dr. Cole, I began discussing my experiences with my friends. I wondered if there was a better, less expensive, and more readily available way to help students with reading difficulties.

In 1982, I was teaching Sunday school at the First Parish Church in Lincoln. My fellow teacher was Dr. Littleton Meeks, known as "Lit," a radio astronomer working at Lincoln Lab who had a great deal of experience using computers in his work. He was a scientist who had spent his professional life exploring, questioning, and finding answers to a variety of problems, and I hoped he might give me some advice. One Sunday I told him about the flashcard event I had experienced at the Greenwood School and asked for his help. Lit thought it was an interesting problem.

I explained to him that I had learned from Dr. Cole that the English language was not easy to master. Our son had the advantage of an O-G tutor and an expensive special school away from home where learning was practiced with tutors on a one-to-one basis. How many children in public schools were deprived of this experience because their teachers did not have time to provide one-on-one practice? Could a computer program do this loosely defined job? Lit thought we should talk to Dr. Cole to get his opinion.

Dr. Cole was eager to discuss this with us and readily agreed to help. His main concern was that the sound of a computer's voice as it pronounced vowels, consonants, and words must be of the highest fidelity and clarity.

Talking with Dr. Cole and Lit about computers and reading was the first step toward creating Lexia Learning Systems. With Dr. Cole's encouragement, the three of us began our adventure. Our goal was to replicate the one-on-one O-G tutorial and flashcards in computer-based programs, including repetitions and automatic branching. This was the beginning. My mission grew from teaching Bo to read to teaching all dyslexic students to read.

If a dyslexic child was fortunate enough to be diagnosed by Dr. Cole and if the parents could afford the tuition at an appropriate school, then the child's prospects for obtaining a good education were excellent. But relatively few children would have that opportunity. I thought again of all the students like Bo who might benefit from moderately priced software. It was a dream that humbled and motivated me.

Our first item of business was to set up a corporation. A corporate structure was necessary for raising capital, for tax purposes when employing people, and for delegation of responsibilities. With the Carroll School so nearby—and so well positioned to help us with their O-G teachers and their deep experience with dyslexic students—we included Carroll in the corporation. Meeks Associates was incorporated in 1982, with 100 shares at $1.00 per share. Lit purchased 60 shares, Dr. Cole 10, Carroll School 10, and I bought 20.

Lit took early retirement from Lincoln Lab. He knew nothing about the acquisition of reading or the O-G code, so the first step was to get him educated. Dr. Cole arranged for office space at the Carroll School and advised Lit to go there and observe how they were teaching reading to dyslexic children. At Carroll, Lit could observe classes and rub shoulders with teachers who knew all about the O-G method.

"I was teaching summer school at the Carroll School when I met Lit Meeks at the copy machine. He told me he was applying for start-up money for teaching O-G on the computer. I questioned him about it and thought it sounded interesting. That's how I got involved with the project at the very beginning."

—SHARON MARSH

The project would need money and equipment. Lit had had a good deal of experience in obtaining grants to finance his own work. He applied for and won a $50,000 Phase I grant from the Small Business Innovation Research Program at the US Department of Health and Human Services. The grant provided seed money for a simple computer and associated equipment that would develop two O-G teaching activities as proof-of-concept demos to support a larger grant.

By the time Dr. Cole retired in 1984, the Cole Center was established at the Carroll School. It would house his handwritten patient records from fifty years at the Language Laboratory and from his private practice. Lit wondered if an analysis of these records would reveal any common diagnostic indicators, which might in turn influence the work he expected to be done by the computer. Could a diagnostic test for dyslexia be developed? If so, the analysis of these records should be done as soon as possible. We raised funds from a number of organizations and individuals to finance this work.[6]

The town of Lincoln was fortunate to have Ken Olsen, the founder of Digital Equipment Corporation (DEC), as a longtime resident. Townspeople often saw him at the grocery store picking out his fruit and snacks. When I served on the Lincoln Conservation Commission, I knew him as a generous contributor to land-conservation projects. I hoped that he might help get our project started with a computer. I called him, described our project, and asked for his help. Mr. Olsen generously agreed to supply us with a Professional DEC 380 microcomputer equipped with database management

---

6. Contributors included the New England branch of the Orton Dyslexia Society, the Francis R. Dewing Foundation, the Laura Stratton Dewey Foundation, the Phineas W. Sprague Memorial Foundation, Thomas H. Adams, Jr., and Robert M. Rosenberg.

software. Lit was instructed to go to the Mill in Maynard and talk to the engineers about his needs. We received not only the computer and software but also a touchscreen and a device that simulated human speech, both of which we could use as we developed the instructional software.

At the Cole Center, we now had Dr. Cole's patient records, the donated computer, and the database management software. With the hiring of Lincoln resident Mary Ann Hales, patient-records analysis was started in July, 1984.

"I saw Lit in Lincoln's grocery store. He asked me if I was interested in a little project. He was seeking someone to work on Dr. Cole's patient records. I signed on as a part-time consultant, and went with Lit to DEC to pick out the computer."

—MARY ANN HALES

The database was designed to hold nearly 2,500 patient records. Before entering all of Dr. Cole's records, letters were sent to his former patients to assess the effect of their specialized education, how far they went with their formal education, and what courses they had found the most difficult. Nothing was found in the data analysis that would point to what our software program should contain.

All that work produced no usable results. Many small businesses that experience such early failure quit at this point. Over the years, Lexia would continue to try new ideas. Some succeeded, some failed, and others nearly compromised our mission.

*"I just wanted to get out of Greenwood and
start a clean slate in a regular school."*

—Bo Lemire

Chapter 9
Finding a High School for Bo

A s the development of our computer-assisted reading program was just
starting, Bo was at Greenwood for the 1982/'83 school year, his second
year at the school.

> "I was feeling more and more confident with my reading ability. I was still
> a slow reader, but at least I understood what I was reading."
>
> —Bo Lemire

Now that Bo's reading-skill remediation was taking root and his reading
ability was expected to be adequate for high school, it was time to make
plans for the fall of 1983. We wondered how his difficult early educational
experience would affect his ability to transition to a regular high school.
Would he be able to leave behind his feelings of inadequacy and the memory
of teachers telling him over and over that he should be doing better? Could
he shake off the feelings of being "dumb"? And how hard would it be for him
to move out of the sheltered environment of the Greenwood School? It might
not be easy just to put all his past experiences behind him and move on with

confidence. I did not know and sadly did not think too much about it. Soon he would have his eighth-grade diploma. What and where was the next step?

Lincoln-Sudbury (L-S) Regional High School was a very good public high school but, with more than 1000 students, we feared it might overwhelm Bo. L-S had been a great place for Bo's sister, Elise, who was unaffected by dyslexia and would go on to graduate from Yale and earn a PhD in English at Rutgers.

> "I did not want to come back to Lincoln to attend Lincoln-Sudbury high school. My elementary school friends would be a year ahead of me. I wanted a clean slate and I could only get that at a school where no one knew me. I had been in a boarding school for two years and had learned the ropes, so finding another boarding school seemed the way to go."
>
> —Bo Lemire

We understood Bo's feelings about entering L-S but were very disappointed that he was not coming home. Greenwood had a short list of private high schools that had successfully enrolled its graduates. We studied the list and set up some visits.

Our first visit was to a Connecticut "boys only" school with heavy emphasis on sports. One of Bo's friends who had graduated from Greenwood the year before was attending this school and met us at the admissions office. Bo was very happy to see him for a brief visit. We went to the parent interview while Bo went into a prospective-student interview.

> "We told our interviewer that Bo was delighted to see one of his Greenwood friends. Over the next hour the interviewer made several disparaging remarks about Bo's friend. He spoke of his lack of preparation, his lack of purpose, and the poor quality of his work.

This was not information he should have shared with us. It made us extremely uncomfortable. We knew before we left the interview that this was not the school for Bo."

—VIRGINIA LEMIRE

We visited two other schools but none seemed the right fit. Perhaps we were just beginning to learn how to be interviewed and how to ask relevant questions. We consulted with Dr. Cole and he suggested that we visit Tabor Academy, in Marion, MA located on Sippican Bay in the southeastern part of the state. All three of us seemed to feel more comfortable at this school. They had a good music department. Bo had a nice singing voice and an interest in musicals. Tabor was co-ed, which made Bo happy.

"I was very glad that there were no Greenwood graduates at Tabor. None of the people we met seemed to know anything about Greenwood. It would allow me to start with a clean slate."

—BO LEMIRE

After some discussion within the family and with Dr. Cole, we applied to Tabor and Bo was accepted.

Graduating with Bo from the Greenwood School in the spring of 1982 were about 12 other boys. They came from Indiana, Connecticut, New York City, New Hampshire, Texas, Oklahoma, and California. Some went on to private schools and some returned to their public schools.

Bob Lemire and Alice Garside

"Alice Garside insisted on rigorous adherence to Orton-Gillingham principles."

—Bob Lemire

Chapter 10
Alice Garside Joins the Team

We put Dr. Cole's records analysis aside. We had been working on our project for two years and did not have much to show for it.

Lit turned his attention to the two O-G activities needed to show proof-of-concept so that he could apply for a larger grant. He studied the O-G method used by the Carroll School teachers. This method combined multisensory techniques with the structure of the English language. Multisensory teaching includes utilizing the three learning pathways: auditory, visual, and kinesthetic. The computer could be programmed to include the auditory and visual features, but the kinesthetic would be harder to implement. In a personal one-on-one tutorial, the student was asked to trace and/or draw letters on paper. It would take some thought to figure how to do that on a computer.

Alice Garside, Dr. Cole's long-time teacher trainer at Massachusetts General Hospital, was now a full-time teacher at the Carroll School. She was widely recognized as a leading O-G trainer. She was about 75 years old at that time but had enormous energy when the subject was teaching dyslexic students. Dr. Cole introduced her to Lit. She quickly saw the potential benefits of computer-aided instruction and wanted to help plan the scope and sequence of the content. Early in 1984, Alice joined the project as a consultant.

116 Chestnut Circle
Lincoln, MA. 01773
December 12, 1984

Dr M.L.Meeks
M. L. Meeks Associates
12 Stonenedge Road
Lincoln, Massacnusetts   01773

Dear Dr Meeks,

Your project interests me because it seems to offer a way to improve instruction of dyslexic pupils as well as to make this instruction available to a large number of pupils in need of nelp.

My chief interest in the field of dyslexia has been in refining teacning techniques for children and adults and in training teachers in the use of these methods.

I would like to serve as a consultant in your project.

Sincerely,

Alice H. Garside

Alice Garside signs on as a consultant

Not only was Alice an expert on O-G training and teaching, but she could provide a bridge between Lit and the Carroll School classes when the project had software to test with actual students. Later, Alice worked out the plan to test the proposed software on Carroll School students, to which Headmistress Margaret Logue gave her blessing.

Alice soon provided Lit with a memo outlining the principles of O-G that should be included in the computer program. This was the first content map for our products. Alice's insistence on rigorous adherence to O-G principles and her iron law that pedagogy ruled technology kept us on track. This is still a critical ideal at Lexia today. Do what works, not just what is easy to make.

Alice strongly emphasized the need for a reliable, computerized test to evaluate the effect of our instructional program. When training teachers, Alice told them that before they started to tutor students, they must first find out what their students already knew, what they had been instructed in, including where practice was needed, and what skills they needed to be taught. Alice suggested to Lit that he contact Dr. Helen Popp, in Maine. She was a retired associate professor at Harvard's Graduate School of Education and a leader in teaching and assessing phonics. Lit sent her his proposal for computer-aided instruction. She responded with a long letter that included many questions and comments. After several more letters and phone calls, a trip was planned to visit her. In June 1987, a friend of Lit's flew him, Alice, and Sharon Marsh to Maine. They met with Dr. Popp for two days of discussions on how to use a computer to teach reading with O-G methods and how to evaluate its effectiveness on student users. Helen had already given the project a good deal of thought.

"Helen Popp convinced Lit that Alice was right about the need for a computerized test to assess students' knowledge and that the test was vital for evaluating the teaching program he intended to produce. Dr. Popp suggested the words and process to be used in the assessment testing and reporting, as well as how to evaluate the students' test performance. We returned from Maine filled with enthusiasm."

-SHARON MARSH

Lit agreed that a computerized evaluation test was of the utmost importance and that it needed to be completed first. Helen had done a good deal of thinking about it. Each test word was to be displayed on the screen and the student would attempt to pronounce the word. Immediately the teacher would hit a key indicating that the response was correct or incorrect. Helen stressed that the time between a word being displayed and the student pronouncing it was critical in determining if the student had mastered the skill being tested, needed practice in the skill, or needed instruction in the skill. The screens were very simple, and there were no graphics, sound, or color involved. By the time the group left Maine, they had the general design of the test well in mind and would call it ASSESS.

"Helen Popp was a very imaginative woman."

−LIT MEEKS

The Carroll School soon moved Lit out of the basement and into a trailer parked behind the school. It was large enough for two desks. Joanne Fraser, the wife of one of Lit's friends, signed on as a part-time programmer.

"When I started working for Lit, in 1984, he knew that sound would be a part of the system. The voice was to be simulated by DECtalk, a gift from DEC's founder, Ken Olsen. The project would use a very limited vocabulary, but nevertheless it had to be clear. Lit and I listened to the various voices that DECtalk produced and chose 'Perfect Paul' as the one with the clearest diction. It was not as clear as Dr. Cole wanted but would work for the prototype."

−JOANNE FRASER

Alice, Joanne, and Lit put together the two prototype exercises required for the Phase II grant application. One was a timed activity called "Bridge" in which the student had to choose the correct vowel before a bridge opened and a bouncing ball fell into a river. The other exercise was an untimed activity for short vowel sounds. Program branching was included to keep students at the same level if they were not successful and branch to the next skill if their work was error-free. Lit made a video tape so that others could easily experience the idea.

Work on Phase I was showing progress in integrating the multisensory functions of sound and touch so that the system could simultaneously display color graphics and respond when particular areas on the screen were touched. Dr. Cole and Lit periodically reviewed the research. Dr. Cole was impressed with the progress and agreed to sign on to the Phase II grant application.

EDWIN M. COLE, M. D.
LANGUAGE DISORDERS UNIT
AMBULATORY CARE CENTER, ROOM 637
15 PARKMAN STREET
BOSTON, MASSACHUSETTS 02114
617-726-2762

December 7, 1984

Dr. M. L. Meeks
Meeks Associates, Inc.
12 Stonehedge
Lincoln, MA 01773

Dear Lit:

I am impressed by what you have demonstrated with the computer system. I can see that it, when programs are fully developed, can go a long way in aiding school systems in identifying children whose learning is retarded due to developmental dyslexia. It is tragic that so many dyslexic students who need special help are overlooked. They lose valuable time in achieving a full education, and not infrequently never get where their potential could bring them if their handicaps were recognized and remediated.

In addition, I can see that there is great potential for developing teaching aids to help schools meet the needs of many dyslexic children. This is a relatively unexplored territory, and I hope you will be able to develop this potential as you continue your work. It has my strong support. I will be pleased to continue as a consultant in Phase II of this program.

Sincerely yours,

Edwin M. Cole, M.D.

EMC:b

50

Dr. Cole's acceptance letter

OMB No. 0925-0226
Expiration Date 2/87

	LEAVE BLANK		
DEPARTMENT OF HEALTH AND HUMAN SERVICES PUBLIC HEALTH SERVICE	TYPE	ACTIVITY	NUMBER
SMALL BUSINESS INNOVATION RESEARCH PROGRAM	REVIEW GROUP		FORMERLY
PHASE II GRANT APPLICATION	COUNCIL/BOARD *(Month/year)*		DATE RECEIVED

1. TITLE OF APPLICATION *(Do not exceed 56 typewriter spaces)*
Computer Diagnosis and Remediation of Dyslexia

2. PRINCIPAL INVESTIGATOR

2a. NAME *(Last, first, middle)*
Meeks, Marion Littleton

2b. SOCIAL SECURITY NO.

2c. POSITION TITLE
President, Meeks Associates, Inc.

2d. MAILING ADDRESS *(Street, city, state, zip code)*
12 Stonehedge/P.O. Box 643
Lincoln, MA 01773

2e. TELEPHONE *(Area code, number and extension)*
(617)259-0093

3. HUMAN SUBJECTS
☐ NO ☒ YES
☐ Exemption # _____
☒ Form HHS 596 enclosed

4. RECOMBINANT DNA
☒ NO ☐ YES

5. VERTEBRATE ANIMALS ☒ NO ☐ YES
If "YES," identify by common names and underline primates.

6. INVENTIONS
☒ NO ☐ YES
OR
☐ Previously reported
☐ Not previously reported

7. SMALL BUSINESS CERTIFICATION
☒ Small business
☐ Minority and disadvantaged
☐ Woman-owned

8. DATES OF ENTIRE PROPOSED PHASE II PERIOD
From: 1 July 1985 Through: 30 June 1987

9. DIRECT COSTS REQUESTED FOR FIRST 12-MONTH BUDGET PERIOD *(from page 4)*
$ 236,250

10. DIRECT COSTS REQUESTED FOR ENTIRE PROPOSED PHASE II PERIOD *(from page 5)*
$ 370,150

11. PERFORMANCE SITES *(Organizations and addresses)*
Meeks Associates, Inc.
12 Stonehedge, Lincoln, MA 01773
Cortical Function Test Laboratory
Massachusetts General Hospital
Boston, MA 02114
The Carroll School
Lincoln, MA 01773

12. APPLICANT ORGANIZATION *(Name, address and congressional district)*
Meeks Associates, Inc.
12 Stonehedge/P.O. Box 643
Lincoln, MA 01773
5th Congressional District

13. ENTITY IDENTIFICATION NUMBER
EIN 1042827970 A1

14. NOTICE OF PROPRIETARY INFORMATION
The information identified by asterisks (*) on pages _____ 30 _____ of this application constitutes trade secrets or information that is commercial or financial and confidential or privileged. It is furnished to the Government in confidence with the understanding that such information shall be used or disclosed only for evaluation of this application; provided that, if a grant is awarded as a result of or in connection with the submission of this application, the Government shall have the right to use or disclose the information herein to the extent provided by law. This restriction does not limit the Government's right to use the information if it is obtained without restriction from another source.

15. DISCLOSURE PERMISSION STATEMENT
If this application does not result in an award, is the Government permitted to disclose the title only of your proposed project, and the name, address and telephone number of the corporate official of your firm, to organizations that may be interested in contacting you for further information or possible investment?
☐ YES ☐ NO

16. OFFICIAL SIGNING FOR APPLICANT ORGANIZATION *(Name, title, address and telephone number)*
M. L. Meeks, President
Meeks Associates, Inc.
12 Stonehedge/P.O. Box 943
Lincoln, MA/(617)259-0093

17. PRINCIPAL INVESTIGATOR ASSURANCE:
I agree to accept responsibility for the scientific conduct of the project and to provide the required progress reports if a grant is awarded as a result of this application. Willful provision of false information is a criminal offense *(U.S. Code, Title 18, Section 1001).*

SIGNATURE OF PERSON NAMED IN 2a. *(In ink. "Per" signature not acceptable)*

DATE
12/14/84

18. CERTIFICATION AND ACCEPTANCE: I certify that the statements herein are true and complete to the best of my knowledge, and accept the obligation to comply with Public Health Service terms and conditions if a grant is awarded as the result of this application. A willfully false certification is a criminal offense *(U.S. Code, Title 18, Section 1001).*

SIGNATURE OF PERSON NAMED IN 16. *(In ink. "Per" signature not acceptable)*

DATE
12/14/84

PHS 6246-2 (Rev. 9/84) PAGE 1

Phase II grant application

*"It took several weeks to research, prepare,
and send off the grant application. Then we
crossed our fingers."*

−Lit Meeks

Chapter 11
Phase II Grant Application

L it wrote the Phase II grant application with a requested budget of about $600,000. The grant application stated: "We propose to develop computer procedures for the early diagnosis and remediation of dyslexia…. This system incorporates animated color graphics, voice response, and a touch-sensitive display. A child using the system will receive verbal instructions while watching a video display and will respond by touching various images on the screen. We propose also to develop techniques whereby a child can be instructed in cursive handwriting with the aid of a graphics tablet…. A multisensory computer system and those procedures that have proven effective will be marketed in Phase III."[7] The goal of the grant was to create a "commercial, marketable product for dyslexic children."[8] Its proposed schedule was July 1985 through June 1987.

---

7. Department of Health and Human Services, Public Health Service, Small Business Innovation Research Program, Phase II Grant Application entitled "Computer Diagnosis and Remediation of Dyslexia," 1984.
8. Ibid.

PRINCIPAL INVESTIGATOR ___M. L. Meeks___

ABSTRACT OF RESEARCH PLAN

NAME, ADDRESS AND TELEPHONE NUMBER OF APPLICANT ORGANIZATION

Meeks Associates, Inc.
12 Stonehedge/P.O. Box 643
Lincoln, MA 01773
(617)259-0093

YEAR FIRM FOUNDED	NO. OF EMPLOYEES *(include all affiliates)*
1982	5

TITLE OF APPLICATION

Computer Diagnosis and Remediation of Dyslexia

KEY PROFESSIONAL PERSONNEL ENGAGED ON PROJECT

NAME	POSITION TITLE	ORGANIZATION
M. L. Meeks	President	Meeks Associates, Inc.
K. L. Gregson	Software Engineer	" " "
Joanne Frazer	Analyst/Programmer	" " "
Edwin M. Cole, M.D.	Neurologist	Mass. General Hospital
Mary M. Chatillon	Director of Reading	" " "
Elizabeth E. White	Head, Cortical Fcn. Lab.	" " "
Robert A. Lemire	President	Lemire and Co.,Inc.
Herman T. Epstein	Professor, Biophysics	Brandeis University
Philip W. Lavori	Assistant Professor	Harvard Medical School

ABSTRACT OF RESEARCH PLAN: State the application's long-term objectives and specific aims, making reference to the health-relatedness of the project, describe concisely the methodology for achieving these goals, and discuss the potential of the research for technological innovation and commercial application. Avoid summaries of past accomplishments and the use of the first person.

The abstract is meant to serve as a succinct and accurate description of the proposed work when separated from the application. Since abstracts of funded applications may be published by the Federal Government, do not include proprietary information. DO NOT EXCEED 200 WORDS.

We propose to develop computer procedures for the early diagnosis and remediation of dyslexia us an augmented version of the multi-sensory system developed in Phase I. This system incorporates animated color graphics, voice response, and a touch-sensitive display. A child using the system will receive verbal instructions while watching a video display and will respond by touching various images on the screen. We propose also to develop techniques whereby a child can be instructed in cursive handwriting with the aid of a graphics tablet. In Phase II we will develop a series of system procedures to provide tools for teachers and psychologists in testing and instructing dyslexic children. These procedures will be developed with the guidance of consultants from the Massachusetts General Hospital and the Carroll School in Lincoln, Mass. Following development of these procedures we will evaluate their effectiveness in trials at these institutions. A multi-sensory computer system and those procedures which have proven effective will be marketed in Phase III

Provide key words *(8 maximum)* to identify the *research or technology* and/or *potential commercial applications*.

dyslexia, language disability, computer, graphics tablet, cursive handwriting

PHS 6246-2 (Rev. 9/84)   PAGE 2

Phase II grant application

The grant's $118,000 proposed budget for equipment included a $17,800 Digital Equipment Corporation PRO 380 development computer with DECtouch, a color touchscreen video monitor. Also included were six classroom computers for student testing, each with a 10MB disk, DECtalk, and DECtouch, priced at $16,740 apiece. In addition to overhead and other miscellaneous expenses, the proposed budget included support for five part-time employees and eight consultants for the two-year period 1985 to 1987.

Proposed Phase II budget

DIGITAL EQUIPMENT CORPORATION   BUSINESS CENTER

If this is a quotation, it shall remain firm for 14 days from the below date unless modified in writing by DIGITAL prior to our acceptance of your contract offer. A quotation is subject to credit approval and is governed by the terms and conditions appearing on the reverse side of this form and/or the terms as noted below.

Please fill in all requested information and forward this form to your Digital Business Center as your contract offer. Insurance will be provided on equipment while in transit, and a charge of $.50 per $100.00 of equipment valuation will be made unless insurance is declined below. You should not execute a contract offer unless the applicable set of terms and conditions is attached to the quotation or unless a discount agreement number is filled in below.

1*Standard terms and conditions of sale	3*Applicable software product service agreement	5*Discount agreement between purchaser and Digital as filled in above
2*Software Professional Services terms and conditions	4*Applicable field service agreement	6*Educational Services terms and conditions

ASSIGNMENT OF SOFTWARE LICENSE AGREEMENT

Purchaser is a bank, financing or leasing company and has entered into this agreement to finance the acquisition of Products, including Software Products, by a third party (hereinafter "User"). Purchaser has received a Software Product License in accordance with the terms set forth in that License.

Purchaser hereby assigns all rights under such License to User, and User agrees to carry out all the obligations set forth in such License.

PURCHASER	DATE	END USER	DATE
Salesperson	Date	By signing this order, I acknowledge that I have read and understand the applicable terms and conditions.	
		Purchaser	Date
Digital's Computer Store Manager	Date		
		This order is executed by an authorized representative of Purchaser.	

QUOTE NO.	DISC. NO.	STORE NO.	PURCHASE ORDER NO.	DATE
0162				07-DEC-84

SOLD TO: 9999
BURLINGTON BUSINESS CTR.
MARKETPLACE PLAZA
BURLINGTON   MA   01803
617-229-8890

SHIP TO: 9999   SLSPSN #09
BURLINGTON BUSINESS CTR.
MARKETPLACE PLAZA
BURLINGTON   MA   01803
617-229-8890

		DESCRIPTION	QTY	UNIT PRICE	NET AMOUNT	TC
06	PC350D2	LARGE SYS UNIT- P/OS LIC-PC350	1	4,025.00	24,150.00	
06	VC241A	PC300 EXT BIT MAP GRAPH OPT	1	895.00	5,370.00	
06	PC3K1DA	COUNTRY KIT (US ONLY)-PC300	1	245.00	1,470.00	
06	RCD51A	10MB.DISK PC350	1	2,400.00	14,400.00	
06	DTC11B	PRO350 TMS VOICE PREREQ:DTC11A	1	295.00	1,770.00	
06	DTC11A	PRO350 PHONE MGT SYSTEM H/W	1	895.00	5,370.00	
06	QBA45A3	COMM DISKETTE - PC300	1	195.00	1,170.00	
06	QBA02A3	P/OS HARD DISK PC300	1	495.00	2,970.00	
06	DTC01AA	DECTALK ALL PC'S	1	4,000.00	24,000.00	
06	BEST	DESTINATION CHARGES	4	80.00	480.00	NT
06	VRTS1A	Color/Touch Video Monitor		3,295.00	19,770.00	

NOTES:

SUB TOTAL	100,920.00 ~~81,150.00~~
TAXES	~~4,033.50~~
	100,920.00
TOTAL	~~85,183.50~~

PAYMENT:
CASH IN ADVANCE XXXXXXXXXX
CREDIT APPLIED FOR

TRANSACTION: QUOTE

CHECK/CREDIT CARD NO.:

FINANCE
TAXABLE
NON-TAXABLE

If this purchase is not a cash or credit card purchase and if it is financed, Purchaser and the financing company must sign DIGITAL's Assignment of Purchase/Sale Agreement. The Assignment of Software License Agreement above does not apply.

TAX EXEMPT NO.:

INSURANCE   MAINTENANCE CONTRACT

INITIAL IF DECLINED   INITIAL IF DECLINED

EN-01047-13-REVC(23E)

7

Proposed Phase II budget

In the grant proposal, I signed on as an unpaid business consultant who was to ensure that Phase II would result in marketable products. The proposal read: "Robert Lemire will head the market research effort that we will conduct during Phase II with funding independent of this grant."[9]

The grant application was sent and we waited for a response.

LEMIRE & COMPANY, INC.
INVESTMENT COUNSEL
P.O. BOX 303, CODMAN ROAD, LINCOLN, MASS. 01773
617/259-0822

December 6, 1984

Dr. M. L. Meeks
Meeks Associates
12 Stonehedge
Lincoln, MA 01773

Dear Dr. Meeks,

I was greatly impressed by the product demonstration of what you have been able to achieve under your initial National Institute of Health SBIR grant. The hardware/software package that you have developed to date clearly demonstrates the feasibility of computer-assisted diagnosis, remediation, and testing of dyslexics.

This is to assure you that I am personally committed to continue serving as a consultant for the work proposed in your follow-up grant application. Moreover, I look forward to working with you as an officer and shareholder of M. L. Meeks Associates Inc., in the marshalling of resources necessary to realize the commercial potential of the products expected to result from the SBIR grants.

It is hard to believe the broad scientific, educational, and commercial interest and support that have been already attracted to your approach for bringing to bear upon the problems associated with dyslexia the full power of new high technology. Based on results to date, I have every confidence that we will succeed in bringing forth practical commercial products that will significantly advance the diagnosis and treatment of dyslexia.

Sincerely,

Robert A. Lemire

Bob Lemire continues consulting agreement

9. Ibid

SUMMARY STATEMENT
(Privileged Communication)

Application Number: 2 R44 HD19822-02
Dual Review: HH

SSS U
Review Group: SPECIAL STUDY SECTION
Meeting Date: FEB/MARCH 1985
HLB

Investigator: MEEKS, MARION L
Position:
Degree: PHD

Organization: M I MEEKS ASSOCIATES
City, State: LINCOLN, MASS
Requested Start Date: 07/01/85

Project Title: COMPUTER DIAGNOSIS AND REMEDIATION OF DYSLEXIA

Recommendation: DISAPPROVAL
Priority Score:

Special Note:
45-HS INV, NO ASSUR; NO IRG CONCERNS OR COMMENTS.
10-NO VERTEBRATE ANIMALS INVOLVED;

PROJECT YEAR	DIRECT COSTS REQUESTED	DIRECT COSTS RECOMMENDED	PREVIOUSLY RECOMMENDED	GRANT PERIOD
02	236,250	-		
03	133,900	-		

RESUME: This Phase II project will develop computer procedures for the early diagnosis and remediation of dyslexia as an augmented version of a multi-sensory system developed in Phase I. This is an ambitious project and may benefit from a rededication of effort to the project's primary mission of developing an effective diagnosis and remediation system for dyslexia. The project weaknesses involve concerns related to technical feasibility, the field testing and the scope of the project. Although Dr. Meeks is an able researcher with a distinguished background in research and development there is a noticeable absence of any research expert in the area of reading disability who will be a major and integral part of the translation process of the Orton-Gillingham instructional program. For these reasons disapproval is recommended.

DESCRIPTION: (Principal Investigator's Abstract) The investigators propose to develop computer procedures for the early diagnosis and remediation of dyslexia using an augmented version of the multi-sensory system developed in Phase I. This system incorporates animated color graphics, voice response, and a touch-sensitive display. A child using the system will receive verbal instructions while watching a video display and will respond by touching various images on the screen. We propose also to develop techniques whereby a child can be instructed in cursive handwriting with the aid of a graphics tablet. In Phase II, we will develop a series of system procedures to provide tools for teachers and psychologists in testing and instructing dyslexic children. These procedures will be developed with the guidance of consultants from the Massachusetts General

The Phase II Grant application was rejected.

"The grant application was rejected."

–LIT MEEKS

Chapter 12

Dr. Pamela Hook Is Recruited

We were not happy when the Phase II grant application response arrived. The grant review scored very high numerically but was rejected with critical comments in three areas. First, the reviewers thought the handwriting part of the proposal was not technically feasible given the time allocated. The second negative comment related to Lit's proposal to develop computer procedures for the early diagnosis and remediation of dyslexia: "The diagnosis of Dyslexia remains a complex process, which cannot be accomplished at this time with only one test. Also, the proposed diagnostic test is not currently recognized as an important diagnostic instrument."[10]

We were encouraged by one statement: "This product must be designed for use in public as well as private schools, since there are very few private schools like the Carroll School. With this in mind, the commercial applications are cautiously optimistic...."[11]

The third negative comment said: "There is a noticeable absence of any research expert in the area of reading disability who will be a major and integral part of the translation process of the Orton-Gillingham instructional program.... Involving an individual with credentials in educational

---

10. Application Number 2 B44 BD 19422-02 "Computer Diagnosis and Remediation of Dyslexia," Summary Statement, page 2
11. Ibid.

research as well as in reading disability might strengthen the proposal.[12] Lit was privately told he should find such a person, delete the handwriting proposal, and resubmit the application.

Lit soon became aware of Dr. Pamela Hook, who had a PhD in communication disorders and learning disabilities from Northwestern University. Lit learned that she lived in Texas but was coming to Boston for one year while her husband pursued a graduate degree. Pam did not have a job that year so she agreed to consult with Lit at his office in the trailer at the Carroll School. At the end of the year, Pam returned to Texas but continued to consult by long distance. Her continuing advisory role was crucial.

"I knew the Carroll School because in the early 1970s I taught there as an intern while earning a Master's degree at Harvard in reading. I knew within the first few days of teaching at Carroll that I did not really understand why the children there were struggling so much to learn to read. My knowledge of O-G came from self-learning, as Harvard's graduate courses in teaching reading were less structured than O-G and did not focus specifically on the child having trouble learning to read. This experience led me to pursue a PhD at Northwestern in communication disorders with a special focus on dyslexia."

—PAM HOOK

---

12. Ibid.

"The grant came with no strings attached."

—Lit Meeks

Chapter 13
Phase II Grant Work Begins

The Phase II grant application was resubmitted after removing the handwriting component and adding Dr. Hook as the requested credentialed educational researcher.

We did not have to wait very long before the grant was approved for $500,000 by the National Institutes of Health under their Small Business Innovation Research program. It came with no strings attached: no oversight, no supervision, no final reports required. Lit quickly began looking for a full-time programmer and better office space.

> "I knew Lit from Lincoln Lab, where I was a programmer in his group. He approached me with a job. Being just a couple of years out of college, I was not sure I wanted to work where there was no one my age. But I accepted—and am still here nearly 30 years later."
>
> —Nancy Johnson

Some of us believed that Nancy signed on because she could bring her black lab to work with her. The staff, now consisting of Lit, Joanne Fraser, Sharon Marsh, and Nancy Johnson, needed more space. Soon the company moved to a one-room second-floor office at 160 Lincoln Road, adjacent to Lincoln's commuter rail station. Pam Hook was in Texas and consulted with Lit at a distance, as did Helen Popp from Maine.

Lit Meeks and Nancy Johnson in the Meeks Associates office at 160 Lincoln Road

Nancy and Joanne worked on the DEC microcomputer equipped with Microsoft Disk Operating System (MS-DOS). Sharon and Helen worked steadily on the detailed contents of the ASSESS testing program. This test would evaluate students' knowledge of each of the 44 sounds and 150 letter combinations that make up most of the words in the English language. The program would branch on each student's performance so that the test would continue as long as the student was successful and end when the student's knowledge was lacking. Each student's performance would result in a printed evaluation showing what decoding skills the student knew, what skills needed practice, and what skills needed instruction.

Pam, Alice, and Lit started working on the overall content of the instructional reading program they called "Touch and Learn" (T&L). The goal for T&L was to replicate the one-on-one O-G tutorial as closely as possible. An O-G tutorial ordinarily encompassed a student's hearing, touching/pointing, speaking, and writing the material presented or required. Our system would not be able to do the writing component because our grant work did not include it. Nor could our computer hardware and software "hear" the student—so Lit and Alice discussed the possibility of the computer generating a list of the words the student had seen during any one session or activity level; the student would then be directed to take the list to the teacher so the student's word pronunciation could be tested. The idea was soon dropped because a printer was not required for our product and might not be available in the classroom.

By now we were all feeling good about our work. We were excited about what we were doing and everyone pulled together. We felt we had the energy to solve most any reading problem.

Bo Lemire's Tabor graduation photo

"I did not want to be treated as 'different'
anymore."

—Bo Lemire

Chapter 14
Bo in High School

"In the fall of 1983, when I arrived at Tabor Academy, we discovered that the school had admitted more students than they had rooms and beds. Three other boys and I were assigned rooms at a teacher's large house across Sippican Bay until rooms could be found on the main campus. We ate our meals in the dining hall but every morning and night we sailed across the bay to our 'dorm.' This didn't seem the best way to start at a new school."

—Bo Lemire

Parent/teacher conferences were scheduled regularly at Tabor. Every report we received about Bo included some form of the phrase "he should be doing better." Clearly Bo was showing his intelligence in the classroom, but he was not matching that with his written performance. The education he got at Greenwood was not magic; it would still take time for him to catch up. Bo's English teacher was more blunt than his other teachers. He was not satisfied that Bo was working hard enough on the written work even though he was doing well in class discussions. We asked him if he knew anything about Bo's educational history. "No," he replied, "and I do not want to know." We were surprised, to say the least. Looking back, however, I wonder if that might have been a good thing. Bo wanted a clean slate and to be treated like everyone else.

In fact Bo was not like everyone else. His O-G training at Greenwood had given him knowledge that many others did not have. This was demonstrated to us when we met the faculty member in whose home Bo spent his first weeks at Tabor.

"Sailing back home across the bay one fall evening, I looked up and pointed to the moon. 'Look at the gibbous (with g as in gym) moon,' I said. Bo immediately corrected my pronunciation of the word gibbous (with the g sound like g in give). When we got in the house I looked it up, confirmed that Bo was correct, and complimented him on his knowledge."

—TABOR FACULTY MEMBER

Commenting on this story recently, Bo said that at least he learned some practical uses of the O-G he had been taught.

"I was more confident in my reading after Greenwood. I was glad I could get through the required content in my classes at Tabor and get decent grades without the teachers knowing my history."

—BO LEMIRE

Early in Bo's freshman year, our local physician saw Bo and spoke to us about his back. The doctor thought Bo should be seen by an orthopedic doctor. The back problem was explained to us as a serious upper-back curve resulting in the front side of each vertebra growing less than the back side. We now think the condition is called Scheuermann's disease. The doctor prescribed a full-torso Milwaukee brace that would relieve the pressure and permit normal growth to occur. Bo's brace was terrible but he wore it the prescribed number of months during his first year at Tabor. I think we may have been hypersensitive to any problems with our son and we did not want to see him have more difficulties. Little did we know that the preventative measure of a back brace may have been worse than the original problem, at least in the short term: his spine completely recovered and his back is now straight and strong, but at a huge price to Bo.

"Wearing a back brace as a freshman in high school and living in a dorm with juniors because of overcrowded freshman dorms was about as awful an experience as you can imagine. I hated it. I hated my parents for making me wear it and I hated the doctor who thought it was a good idea. Take a learning-disabled student trying to pass himself off as a normal kid and make him wear a giant metal back brace as he is adjusting to a new school—it was terrible. I could hide the learning disability well because I had lots of practice doing that. It was impossible to hide a fiberglass girdle around my hips with front and back braces that connected to a metal collar around my neck. I could not wear the required shirt and tie. I couldn't even sit down in a regular chair."

—Bo Lemire

As a junior Bo sang the role of Anthony in *Sweeney Todd*. Unfortunately, Virginia and I were scheduled to spend three weeks in China with a Massachusetts Agricultural Delegation, and were thus able to see only the dress rehearsal and cheer Bo on. It was great for us, but years later Bo says he hardly remembers it. There were so many challenging things in his life at Tabor that perhaps this success simply got lost in his memory.

"I had a few things that I was good at while at Tabor. Sports and music were the main ones. Because these activities did not go well together,

> I did not consider myself part of either group. I did catch a lot of grief on the lacrosse field, but I let my stick do the talking."
>
> —Bo Lemire

The parent/teacher conference with Bo's Latin teacher was much different from those with his other teachers. He told us that Bo was good at Latin and suggested he take the second-year class. When we discussed this with Bo, he flatly refused. Since his diagnosis, Bo's mother and I had often advised him to stay away from a foreign language because he had already learned a foreign language and it was English. We still feel guilty for telling him that.

> "Latin was a logical language with consistent rules and few exceptions. It was much easier to learn than English."
>
> —Bo Lemire

In September 1985, Hurricane Gloria was forecast to come directly into Sippican Bay. We learned that Tabor planned to move some students to upper floors in dormitories that might flood. I called the school and asked if we should bring Bo home for the duration of the storm. The man on the phone said the official policy of the school was for local parents to pick up their student before the storm hit. Then he cleared his throat and lowered his voice. He told me I might like to know that if we came to pick up Bo, Bo would probably never forgive us. We wanted Bo to have all the experiences that Tabor had to offer, and so we let him weather the storm at school.

"The hurricane was fantastic! It was a great opportunity to see nature's power. When it was over, there were sailboats littered across the shoreline. Heavy lifting helicopters came in and moved the boats back into the water or to drydock."

—Bo Lemire

I recall that every one of Bo's Tabor roommates left the school or was dismissed. When we dropped Bo off at the beginning of his senior year, the headmaster asked him who he was rooming with. Perhaps he was wondering who would be the next to leave.

"My Tabor roommates were cursed. Not one of them survived to graduation. My early reading difficulties led to other issues and I tended to gravitate toward the rougher kids. Not many of them survived the structure of a boarding school. I consider myself very lucky to have graduated."

—Bo Lemire

During Bo's senior year at Tabor, I took him to visit several colleges. Virginia had done this with our daughter and I was happy to have the opportunity to make these trips with Bo. He had worked very hard to get to this point in his life.

"My grades gradually improved over my four years at Tabor. When I graduated, in 1987, I had managed to rank just inside the top 25% of my senior class."

—Bo Lemire

Where would the next step take him? Bo did not show enthusiasm for any of the schools we visited. Perhaps he thought one college was as good as the next, or maybe he was tired of school and all the extra work he did for his classes. He didn't say. Nevertheless, he applied to a few colleges that we had visited and was accepted at two of them.

Virginia's parents joined us in attending Bo's Tabor graduation. We were all thrilled with his accomplishments.

"In the fall of 1987 I went to Ohio Wesleyan University. It was the farthest away from home of the colleges that accepted me. Again, I knew no current or former students. I was happy to have a clean slate to begin college."

–Bo Lemire

Chapter 15
From Research to Business

Many small tech startups have trouble remaining focused on their goals because making decisions, large and small, constantly challenges adherence to those goals. We were no different. We were entering the educational world, a world that was much different from the general business world that I knew. We were also attempting to enter the educational world with a new kind of instructional technology. There were few computers in schools and little interest in obtaining and using them. A passion for improving the instruction of reading for dyslexic children was all we had. Would that be enough to make us successful?

As the ASSESS product neared completion, Lit needed to make the transition from a research group to a business. This would not be easy: Lit was a scientist and understandably had little interest in business. We certainly had no money for the business people we needed, but if we did not sell anything, we could not go forward and would have to close. Failure was not an option. I agreed to help Lit with some of the business issues and, fortunately, a Lincoln woman who was interested in marketing presented herself to us:

"I heard about Meeks Associates while studying for my Master's degree in business. I was helping my sister install cabinets and met the job foreman, Mott Meeks. Mott suggested that I talk to his dad, Lit Meeks, in Lincoln. I soon became a marketing intern at Meeks Associates."

—JUDITH COOLIDGE JONES

Lit's one-room office seemed to grow smaller with the addition of Judith. The need for more space prompted Meeks Associates to move to a larger space across the railroad tracks, at 11A Lewis Street. The small two-storied building had been a garage, with a cold cement ground floor and a heating system that left much to be desired; but it was in Lincoln and everyone wanted to remain in town. Nancy Johnson could continue to take her dog out for walks and for a swim in a nearby pond.

Design and development continued on the O-G instructional activities in Touch and Learn. Lit recruited his art-school-graduate daughter-in-law, Catherine Meeks, to create the graphics for T&L. Catherine had no computer experience, so Nancy showed her the basics and gave her a list of drawings that needed to be created for the software under development. Sixteen pixels—4 x 4—were the maximum allowed for each icon drawing.

"I used a drawing program called 'Dr. Halo,' along with a tablet and pen hardwired to a PC. Nancy configured and booted the PC each morning before I arrived. At the end of the day, I delivered my results to her on a 5.25-inch floppy disk, as there was no office network. I had a very limited choice of colors with which to make the icons. If the icon was to move, I'd make six to eight drawings representing the progression of movement that Nancy animated on her PC so that it looked like a movie. I had complete discretion over what the drawings looked like. It was so rewarding for me to do meaningful work."

—CATHERINE MEEKS

The Meeks Associates staff had now increased to include Nancy Johnson, full-time programmer; Elaine Briggs, part-time programmer; Judith Coolidge Jones, full-time marketing; Catherine Meeks, part-time graphics programmer; Sharon Marsh, part-time developer of curriculum; and Mary Ann Hales, who was writing the ASSESS manual. Language consultants were Alice Garside and, via mail and phone from Texas, Pamela Hook.

Suddenly, the new and growing Apple family of computers was becoming visible in schools. Our programmers began investigating the development environment on the Apple computer Versions II and IIC. ASSESS screens were black and white and had no sound. How difficult would it be to transfer the software code from the IBM Multi-Color Graphic Array ASSESS program to the Apple computer? We called one of Lit's software-engineer friends for advice. He provided a road map for that effort as well as a strategy for "hiding" the code to keep someone else from using it. Nancy was then able to adapt the PC version of ASSESS software to the Apple II in time for it to be beta tested with the PC version with students at the Carroll School. We also made contact with the superintendent of schools in Wayland, MA who became very interested in the project. In 1988 and 1989, Meeks Associates tested first-graders on PCs at the Sleepy Hollow School in Wayland under its principal, John Talbot.

The staff continued to work on bug fixes and enhancements for ASSESS. By May 1989, the attempts to sell ASSESS to schools had largely failed because of flaws and limitations in the product. Teachers who tried it were in many cases discouraged by the awkwardness of the input, bugs that interrupted the working of the program, and the inability to erase or change data. Trying to sell ASSESS before it was useful to teachers and stable enough to be successful could have been fatal. We learned a hard lesson from this. The product had been enthusiastically received in Wayland and at the Carroll School, but in those installations we had helped them use it. We asked ourselves: should we go to market with the current version of ASSESS? Should we wait until it was a more solid product? Should we develop a home version of ASSESS? Should we get some help from professionals in developing a marketing strategy? Did we have money to do any of these? Were we remaining true to our mission?

A plan was agreed upon to update ASSESS into a robust product, launch a sales campaign, and sell a minimum of 100 units per month to achieve break-even. The order of this work was critical. We could not sell a product that was unreliable.

In addition, we wanted to complete a "pseudo-word" unit in ASSESS to test specifically for decoding abilities. Even at the lowest priority it was judged to be an important addition to ASSESS: it would be used for those students who had exceptional memories and knew several thousand words by sight but could not decode. The pseudo-words would follow O-G rules but were not words these students would ever have seen before, for example, zop and zope. The pseudo-word test turned out to be very useful for the early identification of non-decoding students of whom teachers were not aware.

"I often tested poor readers and although they could correctly pronounce the displayed real words they failed on the pseudo-word section. This proved to teachers and parents alike that many students were not able to decode words. That often opened the door for parents and students to accept that they needed further reading training."

EARL OREMUS, HEADMASTER, MARBURN ACADEMY

My part of the agreed-upon plan was to raise money from private investors to keep the effort going. To start, I raised money from my family before asking friends and associates for investments.

Once the ASSESS software was considered reliable, the manuals printed, an "800" telephone number set up for customer support, and packaging made ready, ASSESS was ready to ship. The printed materials had been outsourced but reproduction of the 5.25-inch floppy disks was not outsourced because the software was still not stable enough to merit reproducing large quantities. Bug fixes would continue to be necessary as customers with different platforms and configurations encountered problems. Duplication of the floppy disks would have to remain an in-house job.

As we struggled with the transition to a business, Lit agreed to bring Dr. Henry Morgan onto the Meeks Associates board to serve as a director with Dr. Cole, Alice Garside, and me. Henry was Dean Emeritus of the School of Management at Boston University, a local entrepreneur, and friend. He had a PhD from MIT and could review the science of what we were trying to do. I had no scientific experience or education but agreed to spend more time at Meeks Associates in an attempt to bring some business practices and vision to the operation.

We were surviving, but it was not a comfortable situation.

The 1989 release package of the ASSESS software.

Meeks Associates Staff in front of 11A Lewis Street ("The Garage"). Back row: Elaine Briggs, Sharon Marsh, Judith Coolidge Jones, Nancy Johnson. Front row: Pam Hook, Lit Meeks, Joanne Fraser

"Once Lit had successfully completed the research, his interest faded."

—Bob Lemire

Chapter 16
Corporate Transition

I stopped by the Meeks Associates office at 11A Lewis Street most mornings when I walked to the coffee shop from my home office on Codman Road. I was not involved in the day-to-day operation but felt the need to show up on a regular basis. That was about to change.

Lit was an able scientist who, understandably, had little experience or interest in managing a company and none whatsoever in marketing. His interests and skills were in doing research and proving that the proposed work could or could not be done. We were, and remain, profoundly indebted to the basic research he accomplished for dyslexic students. Once he had accomplished this research, he started to look for another project.

Lit was still interested and very excited about doing research on handwriting. He had wanted to include handwriting in the Phase II work, but the grant reviewers would not fund it. He applied for and won a $500,000 grant from the National Institutes of Health for handwriting research. I recall him saying that there might be interest from the credit-card industry if his research came out well. Nancy Johnson and Elaine Briggs were part of Lit's new grant and he assigned them to spend 30 percent of their time on the new handwriting project. That meant ASSESS and T&L were going to take even longer to become solid products.

In October 1989, Lit decided he wanted to focus all of his time and work on handwriting and asked to get out of the reading project altogether. Although I was not surprised, I was not prepared for this change. I had raised investment capital for Meeks Associates. If the business collapsed, I would have a lot of explaining to do to my family and friends. We could not let all this important work simply fade away. We could not abandon our mission. We had to carry on without Lit.

Henry Morgan and I worked out a plan to start a new company that would buy Lit's shares for $150,000, payable over time under a royalty agreement. After some discussion and legal advice, we simply changed the name of the company from Meeks Associates to Lexia Learning Systems, Inc.

Lit wished us well and moved on.

"We took the 'dys' out of dyslexia."

– JUDITH COOLIDGE JONES

"Whenever Bob came in, I would meet him at the door, shake his hand, and say, 'Welcome to Lexia Learning Systems' "

– JUDITH COOLIDGE JONES

Chapter 17
Lexia Learning Systems, Inc.

Following Lit's departure, I had to assume responsibility for day-to-day activities as well as planning for the long term with the Board of Directors. That was not exactly what I had signed up for—but I did not think I had a choice. The grants had run out, our first product introduction had just taken place, and we were on our own; I would just have to raise the $9,000 per month required to keep Lexia going until sales could support us. As we looked ahead, we had no idea how long that would take and were mindful of the sizable risk. But the goal of getting these products to those who needed them was all that mattered. That goal would become very difficult to pursue.

The change from Meeks Associates to Lexia Learning Systems was an event for the staff as well as for me. They were much more on their own and were looking to me for leadership.

I helped Judith use her newly acquired business skills to set up business procedures. She took on the administrative side, which included everything except program development, where Nancy was in charge. Judith worked to create marketing materials, including brochures, press releases, and product manuals. We had little idea how to market the products except to attend educational seminars, conferences, and trade shows and place one or two advertisements in trade publications.

"Sharon Marsh and I went to a large convention and sold one copy of ASSESS. We were so excited that we called the office to tell everyone."

– JUDITH COOLIDGE JONES

Soon we were selling a few copies of ASSESS. Nancy took the support calls and occasionally made immediate bug fixes and sent out a modified version that solved the customer's problem. To protect the source code, she took floppies to her home and I took them to mine for safekeeping.

It was not long before Nancy requested help for a formal quality-assurance (QA) check. Debbie Gillespie, the wife of a local software engineer who worked with Virginia, was hired to do QA for ASSESS A and later ASSESS B. She had her own IBM PC, so she could work at home, and in our penny-pinching mode we were happy not to have to purchase another computer.

Although we knew at the time that not many classrooms had computers and not many schools had computer labs, we did not anticipate how hard it would be to sell ASSESS. Teachers were doing whatever reading assessments they did by hand and did not feel a change was necessary. We stressed that the computer test was quicker and more standard for each student, that it generated individual and class reports and saved teachers' time. We felt that we finally had a product that would make a difference in student lives—but putting it in the hands of teachers was a problem we would have to solve soon. We still had a lot to learn about the school market.

In September 1990, Pam Hook returned to Boston. She had the credentials Lexia needed, so we hired her as a part-time consultant. Her responsibilities were (i) to take charge of new program development, including writing protocols for validity and reliability studies, (ii) to support sales with local in-service workshops and attendance at conferences, and (iii) to write grant applications. Pam was just the right person for the job and we were thrilled to have her.

I stopped by the office one night when I saw the lights on and discovered the language consultants working on something. When I asked them what they were doing, Alice Garside replied: "We're plotting a revolution!" It would be some time before I learned how much of a revolution it would take, and continues to take, to change how reading is taught in our nation's schools. Clearly Lexia was on the cutting edge of creating computer-based phonics reading programs for schools and individuals, and computers were just starting to be available in school classrooms and computer labs. Teachers of that day were not technology-users and were not interested in having a tool in their classroom that they did not fully understand. I am sorry to say that this problem has not yet completely gone away.

While we were trying to learn about the school market, we needed cash to keep the company afloat. Henry Morgan put in some much-needed dollars. I purchased stock for Virginia and our two children. When Lexia's financial situation grew critical I approached members of my extended family. I went to local people who knew me and my record in town affairs. Sometimes the people I approached referred me to their own family members who were teachers, educators, or otherwise interested in education. I could not tell any of them when they might be able to get their money out. Between 1984 and 1994, I raised about $700,000 in capital from about 25 individuals and trusts. Many of them purchased shares on more than one occasion because they believed in what we were doing. Share prices started at $10 and rose over time.

> "We nervously joked about people crossing the street to walk on the other side rather than meet up with Bob and be asked for investment money."
>
> —VIRGINIA LEMIRE

For several years, cash was so tight that I would dread the days when Lexia did not have enough funds to meet payroll. Sometimes the payroll came from my own money and that of my extended family. Sometimes an angel investor came forward in the nick of time. I continued to be an investment advisor but stopped consulting on land-use issues because Lexia needed my time. It was 18 years before I could be added to the payroll.

We retained Mary Ann Hales at Lexia for as long as there was money to pay her. When the money ran out, she got an IOU for her last weeks. The IOU was eventually converted to stock.

> "Bob seemed to be calling me at work every other week to request my OK to lend Lexia some of our money for the payroll. When was Lexia going to be self-sufficient? It was touch-and-go for so long. We could not buy more stock because we could not undertake the risk. Eventually we were repaid, but I know it caused Bob enormous stress."
>
> —VIRGINIA LEMIRE

The Lexia Learning Systems, Inc., First Report to Shareholders, dated February 4, 1990, stated, "Lexia Learning Systems, Inc. (LLSI) began operating as a separate entity on October 1, 1989. LLSI was created to continue development and commercialization of the reading skill diagnostic and exercise programs pioneered by Meeks Associates, Inc., with grant funding from the National Institutes of Health."[13] The Report went on to say that ASSESS had achieved sales of $2,672 while expenses totaled $34,999 for the quarter ended December 31, 1989. During this period LLSI took in $140,100 in private capital investments. Although it had then been seven years since the company's founding, private capital would continue to be essential for Lexia's survival.

The Report further stated that some $100,000 to $150,000 of new capital would be required during 1990. Lexia's Annual Report for 1990 stated, "Capital infusions of $186,000 including $31,000 in loans from officers

---

13. Lexia Learning Systems, Inc., First Report to Shareholders, February 4, 1990.

['officers' meant me] were required to get us through 1990." Such was the shaky financial condition of Lexia.

The financial situation did not improve much the next year. The 1991 Annual Report revealed that Lexia's sales that year were $27,076 with a net operating loss of $189,985. We had $197,000 of capital infusions that year including $27,000 in loans from officers (me, again) in 1991. That year Lexia won the "Award of Merit from Curriculum Products" and was included in "Districts Choice – Top 100 Products of the Year." Many educational awards followed over the years, and although awards always boosted our morale, I'm unsure how they translated to our bottom line. At least it publicized the Lexia name.

Eventually I was repaid the payroll money that had come out of my own pocket. I am grateful to Virginia for sharing Lexia's mission with me and for her years of work in the computing industry so that we could maintain some financial stability for our family. There were so many times when both Virginia and I were just weary of the whole thing.

But what would I be looking back on with such satisfaction if it were not for Lexia?

SOFTWARE FOR BEGINNING READERS

LEXIA
LEARNING
SYSTEMS, INC.

Lexia's Touch and Learn Reading Software package

> *" 'Mastery before advancement' was the key to Lexia's individual student learning success. "*
>
> —Bob Lemire

Chapter 18
Touch and Learn

The essential ingredients of Touch and Learn were (i) the skills to be taught and practiced, (ii) the activities designed to teach the skills, (iii) the words required for each activity, and (iv) the branching needed to repeat the activity or advance to the next activity based on the student's performance. Advancement was based on sequential skill mastery. This "mastery before advancement" concept was the essential discipline in Touch and Learn (and all Lexia's products), and technology is what allowed this learning process to be individualized—one-on-one. This is what made Lexia different from everyone else.

Pamela Hook took charge of the detailed content of T&L, with direct help from Alice Garside, Sharon Marsh, and Barbara Porter. They knew what curriculum they wanted to include and what they wanted to accomplish, but how to do it was challenging. As the effort progressed, they agreed on the overall contents for T&L. Then the team divided the content into three groups, Levels I, II, and III. Detailed design was done by skill and then by level, so that the programmers could separately implement, test, and release each Level for sale as it was completed.

As the detailed design work on T&L continued, the target hardware platform and software tools had to be defined. Choices of computers were rapidly increasing in number; which computer system should be our chosen platform? There were IBM PC, Tandy, IBM-compatibles, and Commodore, among others. Some of these ran MS-DOS but there were other disk operating systems used as well.

Even though the presence of Apple in schools was increasing, we decided to stay with IBM because we knew this computer. We had built the prototype on it with the MS-DOS operating system. The Apple version would have to wait.

As I wrote earlier, the O-G method used auditory, visual, and kinesthetic methods and that the auditory and visual were easy to implement on the computer. For the kinesthetic method, Lit had wanted to add an electronic tablet, for tracing letters, etc., but the grantors did not support that lengthy work. Lit had fulfilled a small part of the method by requiring the student to select an object, letter, or word on the screen by "touching" it. We initially used a very expensive touchscreen that could be affixed directly to the monitor with Velcro. A less-expensive concept used computer keyboard arrow keys so students could move a pointer to their choice, but it seemed too difficult to teach very young students how to use the keyboard for that purpose. We eventually found a less-expensive touchscreen from Edmark. Each one had to be calibrated to each monitor. It was time-consuming to set up and expensive for the customer to purchase.

As each activity was viewed on the screen by the designers, programmers, and language consultants, changes were often needed. The activity on the screen sometimes differed from how it had been imagined or intended, and seeing students use the activities during beta tests often led to changes as well.

Design work continued as products were tweaked and updated. Over the years, re-design remained an ongoing activity as Lexia products were extended and enhanced for educational, software, and/or hardware reasons. Nancy Johnson wrote the software engine in the C programming language. Her remarkable work remained as the basic engine for our instructional products from the 1980s until the 2013 release of Lexia Reading Core5.

Input from an electronic mouse became available when the IBM Model 25 was introduced, in the late 1980s, and we wondered if the mouse would work with our system. Could young students master a mouse? At that time most students had never even seen one—but we were surprised at how quickly children adapted to this new input device. It is interesting to note that Lexia started with a touchscreen, went to the mouse, and has returned to optional input from a touchscreen on a tablet.

By 1992, our programmers had expanded T&L to run on the IBM Model 25, which brought a multi-color graphics array, 3.5-inch floppies, digitally recorded natural speech, and the mouse. We could now drop the expensive touchscreen from our hardware requirements. Lexia products were also adapted to run on Novell networks, which many schools had implemented. Over time we put significant effort into supporting third-party software and various types of hardware, many of which later disappeared from the mainline of development. Just like today, it was difficult to discern exactly where technologies would go next.

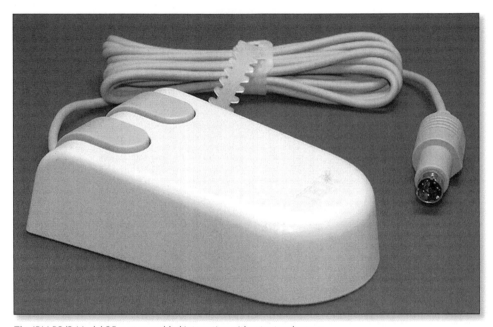

The IBM PS/2 Model 25 mouse enabled interaction without a touchscreen

Lexia "Touch and Learn" Components

Each level covers increasingly difficult phonic elements and reading skills.
Programs available individually or by level. New programs will be forthcoming.

LEXIA A·I
Consonants &
Short Vowels

$250 level

Bridge
• select the correct short
  vowel for dictated words
19 branching units - $50

Change
• manipulate short vowels
  and consonants in 3-letter
  words
7 branching units - $50

Consonant Castle
• select the correct
  consonant for
  dictated words
7 branching units - $50

Sort bdp
• visually sort b's, d's
  or p's
9 branching units - $50

Touch and Listen
• match short vowel letters,
  pictures, and sounds
9 branching units - $50

Touch and Paint*
• bonus: design a picture
  with shapes and colors

LEXIA A·II
Short Vowels &
Long Vowels

$250 level

ABC Race
• alphabetize letters
7 branching units - $50

Balloons
• select and sort words with
  short vowels or long vowels
  (silent-e)
19 branching units - $50

Match It!
• match pictures and phrases
  containing short vowel and
  long vowel (silent-e) words
7 branching units - $50

Score
• choose the correct vowels
  to form short vowel and
  long vowel (silent-e) words
7 branching units - $50

Super Change
• manipulate consonants and
  vowels in short and long
  vowel (silent-e) words
9 branching units - $50

Word Play*
• bonus: create pictures
  by combining words and phrases

LEXIA A·III
Two-syllable Words, Long
Vowel Review, Vowel-R, &
Vowel Combinations
$250 level

Alpha Rocket
• alphabetize words
10 branching units - $50

Elevator
• manipulate syllables to
  create 2-syllable words in
  isolation and in context
12 branching units - $50

Pirate Ship
• select the correct
  vowel combinations
  for dictated words
13 branching units - $50

Train
• identify dictated
  short vowel and long
  vowel (silent-e) words
14 branching units - $50

Word Stairs
• choose consonants and
  vowel-r combinations to
  form words in isolation
  and in context
7 branching units - $50

Pipe Dreams*
• bonus: arrange pipes
  to complete a puzzle

*free with purchase of one or more of the Lexia "Touch and Learn" programs

Hardware Requirements
• PC/AT (IBM 80286 compatible)
• 20-40 meg. hard disk system with 1 floppy drive
• 640K of memory
• Color monitor with EGA or VGA graphics adaptor
• Edmark® TouchWindow or Elographics® Touch Screen or Mouse
• Dialogic® or Digispeech ® Voice Card
• Speaker and/or Headphones (with volume control)
• Optional Dot-Matrix Printer (Reports can be viewed on screen)

Orders & Information: 617/259-8752

© Lexia Learning Systems, Inc. • 11a Lewis Street • P.O. Box 466 • Lincoln, MA 01773

Early Touch and Learn brochure

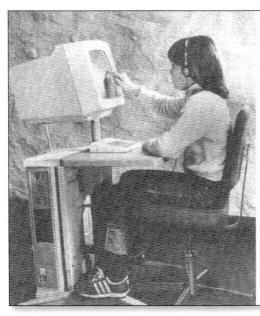

"Touch and Learn"

from
Lexia
Learning
Systems,
Inc.

Using early touchscreen

Minimum Hardware Requirements		
	Macintosh	**IBM/PC Compatibles**
Computer	LC-II or better	AT or better
Color Monitor	12" screen	MCGA, EGA graphics screen or better
Operating System	System 6.07 or greater	DOS 2.0 or greater
Memory	4 MB RAM	640K RAM
Available Hard Disk Space	25 MB	13 MB
Peripherals	Mouse (standard)	Digispeech Audio Adapter Unit* (model # 301PE or DS201A)
		Mouse
	Printer (optional)	Printer (optional)

Touch and Learn platform requirements

Judith Coolidge Jones

> *"Beta testing turned up things we had not anticipated."*
>
> —NANCY JOHNSON

Chapter 19
Beta Testing

W hen beta testing for T&L began at the Carroll School, Nancy Johnson went to observe dyslexic children using the program. One day she watched a 10-year-old boy test the Elevator activity: when he was presented with the first syllable of a word located in a building "elevator" and was asked to find the second syllable on one of the "floors" of the building, the elevator rose to the chosen floor to form a word. The following illustration is what she saw on his screen.

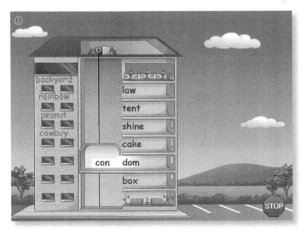

Elevator Activity

"I was shocked to see this possibility on the screen! The student knew 'condom' was a word. I was trying to keep my face straight when the teacher asked the student to go ahead and find another word. I went back to the office and asked the team to review everything to be sure we didn't see something like that again."

–NANCY JOHNSON

The next Touch and Learn beta testing site was in a special education (SpEd) classroom at the Sleepy Hollow School in Wayland, MA. The arrival of Lexia's software at the school proved interesting not only for the SpEd teachers but for all the teachers who saw it. They were very impressed with the program's functionality and told us that if they had Lexia's Touch and Learn in their regular classrooms, perhaps they would not send so many students to SpEd for reading help. Although we had considered placing Touch and Learn in regular classrooms, we were surprised to hear this request from classroom teachers. We knew that the majority of students sent to Special Ed for reading help were not necessarily dyslexic; rather, they were students who just needed more instruction and practice than they could get in the regular classroom. The Wayland teachers indicated that Touch and Learn could provide that instruction and practice right in their classrooms. In hindsight, perhaps we should not initially have focused only on "disabled" students.

We also tested the software in regular classes in our Lincoln elementary school. Again, feedback from the school proved important. The students loved the program and begged each morning to use it. One teacher took her students' Lexia scores home with her every night and made a bar graph so the students could see their progress the next morning when she posted it on the wall. The students loved the graph and so did we. Lexia responded by building the bar graphs (ladders) into the software so each student could see his or her progress every time they signed on and when they finished an activity. This was a significant addition to the students' experience using Lexia. The ability to see their own progress and how far they needed to go to finish a level is still important today.

Touch and Learn included reward modules that were automatically initiated when the student finished a unit. Touch and Paint was one that allowed the student to use a free-form simulated "paint brush" to create a drawing. Water Pipe was another reward the students loved. It allowed a student to connect water pipes in any number of ways, with the sound of rushing water generated when the pipe was fully connected. Students loved these modules at least in part because there were not many computer games at that time. When reward modules were no longer part of the product, we packaged them for individual sale but they failed to generate much interest.

The 44 Sounds in the English Language

5 Short-Vowel Sounds	18 Consonant Sounds	7 Digraphs
short / / in **apple**	/b/ in **b**at	/ch/ in **ch**in
short / / in **e**lephant	/k/ in **c**at and **k**ite	/sh/ in **sh**ip
short / / in **i**gloo	/d/ in **d**og	unvoiced /th/ in **th**in
short / / in **o**ctopus	/f/ in **f**an	voiced /th/ in **th**is
short / / in **u**mbrella	/g/ in **g**oat	/hw/ in **wh**ip *
	/h/ in **h**at	/ng/ in si**ng**
	/j/ in **j**am	/nk/ in si**nk**
	/l/ in **l**ip	
	/m/ in **m**ap	* (**wh** is pronounced /**w**/ in some areas)
	/n/ in **n**est	
	/p/ in **p**ig	
	/r/ in **r**at	
	/s/ in **s**un	
	/t/ in **t**op	
	/v/ in **v**an	
	/w/ in **w**ig	
	/y/ in **y**ell	
	/z/ in **z**ip	
6 Long-Vowel Sounds	**3 *r*-Controlled Vowel Sounds**	**Diphthongs and Other Special Sounds**
long / / in **ca**ke	/ur/ in **f**e**r**n, b**ir**d, and h**ur**t	/oi/ in **oi**l and b**oy**
long / / in f**ee**t	/ar/ in p**ar**k	/ow/ in **ow**l and **ou**ch
long / / in p**ie**	/or/ in f**or**k	short / / in c**oo**k and p**u**ll
long / / in b**oa**t		/aw/ in j**aw** and h**au**l
long / / (yoo) in m**u**le		/zh/ in televi**si**on
long / / in fl**ew**		

84

"I knew my father was getting more involved in this company he called 'Lexia.' I can remember hearing him asking anyone who would listen: 'Do you have any idea how many sounds there are in the English language? Forty-four! Do you have any idea how many letters and letter combinations make those 44 sounds? One hundred and fifty!' His passion for this mission and developing business was amazing. I was doing my own thing, and stayed away from the business."

—Bo Lemire

Chapter 20
Bo in College

We saw and heard little from Bo during his college years. He had drifted further away each year he had been away from home. We visited him in his Ohio college town, where Virginia was reminded of the main street of her hometown in Iowa. The buildings were from another era and were beautifully maintained.

Virginia's sister and her husband, parents of at least one undiagnosed dyslexic child, attended Bo's college graduation with us, in 1991. Bo had completed his degree in four years. Arthur Ashe was the graduation speaker. A few years later, Virginia and I saw a large stack of old Life magazines for sale at an outdoor fair. We found the issue corresponding to Bo's birth week: incredibly, Arthur Ashe was on the cover.

"My father would talk about Lexia when I was home from school but I had absolutely no interest in it. I stayed as far away from it as possible. I did not want to be reminded that I had been a 'disabled student'; I wanted to distance myself from that past. I had been taught how to read with O-G and wanted to forget about all that."

—Bo Lemire

"The Nation's Report Card woke me up."

—Bob Lemire

Chapter 21
Not Just for Dyslexic Students

In 1992, it came to my attention that the National Assessment of Educational Progress (NAEP) had just issued *The 1992 Nation's Report Card*. This biannual report revealed that 40% of fourth-graders could not read at grade level. Not only was this result shocking, but I realized that the asessment mode it had utilized was essentially a whole-language test. It had been derived from a represetative multiple-choice test of reading comprehension of two passages at the fourth-grade level. It did not directly test decoding skills, which led me to think that the real fourth-grade reading levels were likely even worse than the report card indicated.

It was appalling to think that the reading skills in this country were so poor. Many of these students would simply fail to learn to read in our nation's schools. Large numbers would drop out when they were old enough to leave school lawfully. I saw this as a threat to the nation's future well-being.

The Report was my wake-up call to the surprising fact that reading-acquisition failure was not just a dyslexic phenomenon but also a serious problem for many beginning readers. I had seen it in Wayland. The Sleepy Hollow School teachers were right when they said T&L should be in their regular classrooms.

These facts transformed my own and Lexia's mission from serving only dys lexic students to serving every beginning and struggling reader. Lexia's reading programs would benefit all students, so that fewer would have reading problems in higher grades and fewer would drop out. Teaching reading to all students in a structured way could vastly improve the Nation's Report Card. Lexia reaffirmed its resolve to keep developing and selling its products no matter the effort. We now needed to expand our focus and mission to serve every beginning as well as dyslexic English-language reader. This became Lexia's new challenge.

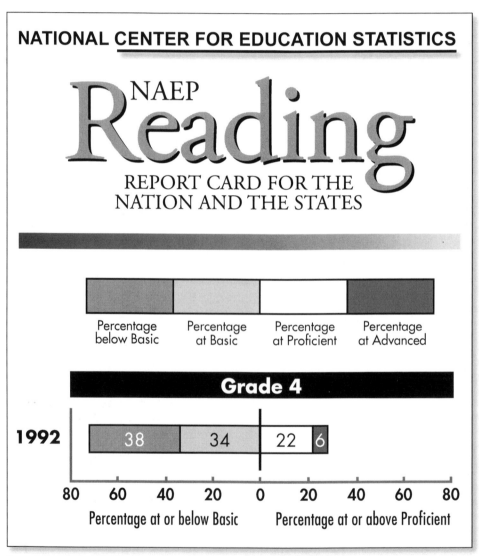

Percentage of students within each achievement level range for the nation in 1992. *Nation's Report Card*

"The fidelity of Lexia's sound is essential."

–Dr. Edwin Cole

Chapter 22
The Voice of Lexia

P roviding the required sounds for Touch and Learn was challenging. We used computer cards from Digispeech and later from Sound Blaster to deliver sound to the student. The sound card was installed on the computer's motherboard and produced sound via headphones or external speakers. Using external speakers was workable in a classroom situation where one or two computers might be running T&L at the same time. If the Lexia software was deployed on multiple computers sited next to one another in a computer lab, the students needed headphones to isolate themselves from the sound of other computers running Touch and Learn. Shared headphones brought to light a different problem, as schools were reluctant to use communal headphones for fear of transmission of head lice. Some schools requisitioned a pair of headphones for each student while others used disinfecting tissues to clean them between users.

In order for the Digispeech and later Sound Blaster cards to deliver the sound, the sounds first had to be recorded and digitized. Our developers initially recorded all the required sounds in our offices, including all the words used in the activities and in the instructions. This was a difficult task because our office was still located by the Lincoln commuter rail station, and each time a train came close they had to stop recording. These recordings were usable for development and testing, but higher-quality recordings were needed for the product. Developers tried recording in a sound room at Massachusetts General Hospital but again the words were not sufficiently clear. The voice needed to be devoid of any accent and the speaker's tone and diction needed to be perfect, especially for the vowels. Although we remembered well Dr. Cole's early insistence on clarity of sound, we had no budget for anything better than what we had.

In 1997, funds became available and Audio Link in Watertown, MA was retained to find the right voice and do the recording for our product. We recently interviewed owner Steve Olenick and were pleased to hear some of the first Lexia recordings from his vast archives.

> "Improved technology has resulted in much smaller sound files and much higher quality since we first recorded for Lexia, in 1997."
>
> —STEVE OLENICK

When Audio Link's voice auditions were complete, Pam Hook and other Lexia language consultants reviewed the recordings and chose one man, Tom Glynn, and one woman, Mary Ellen Whitaker, to be the voices of Lexia products. At first Lexia used both voices, but the female voice was preferred to the extent that the male voice was dropped. Mary Ellen has been the sole "voice" of Lexia's American products since 1997.[14] Pam Hook was in charge of the Lexia voice and supervised each recording session from the control room at Audio Link. Every sound had to meet Pam's extraordinarily high standards for clarity and were often recorded more than once so the Lexia consultants could listen and choose the clearest one. In a few instances, the consultants coached the "voice" to perfect enunciation.

> "I learned so much from my experience with the Lexia recordings. I felt I was given a course in pronunciation. Learning how to sound out words has helped me so much in my other jobs."
>
> — MARY ELLEN WHITAKER

---

14. A few years later, when Lexia products began selling well in England and other Commonwealth countries, a British "voice" was used for the UK version.

Voice of Lexia—Mary Ellen Whitaker

Product packaging for Lexia PBR

"It had taken six years to get Touch and Learn to the market."

—BOB LEMIRE

Chapter 23
Product Announcement

In January 1990, Lexia announced the availability of Level I of the Touch and Learn program. Mary Ann Hales bought the first copy to donate to the Lincoln Public Library. Ken Olsen, who had given Meeks Associates its first computer, gave the library a computer to run it. The availability of T&L Levels II and III were announced in March and April of the same year.

Each Touch and Learn order was shipped in a custom box that contained an instruction manual, a quick reference card, and, if all three Levels were ordered, as many as 18 floppy disks (no one remembers exactly how many). When all the floppies were loaded on a customer's computer, T&L could occupy the computer's entire available memory. The touchscreen and sound simulator were shipped directly from their respective vendors.

> "Judith Coolidge Jones and I went to New York City to show Lexia products at an Ivana Trump–hosted luncheon at the Plaza Hotel. To control expenses, we took the subway, pulled the equipment on dollies along city streets, and slept on the floor in a small apartment of one of Judith's college friends."
>
> —SHARON MARSH

At the end of 1990, it was a huge relief to have Lexia T&L Levels I, II, and III as well as ASSESS A and B shipping and bringing in some money. But the office was never quiet, as customers often had technical problems that kept us busy.

"Lexia banked at West Newton Savings Bank, just across the railroad tracks from the office at 11 A Lewis Street. Lexia had made its largest sale ever and Bob had a check for $50,000. Somehow between receiving it in the mail and depositing it in the bank, the check was lost. He had to call the issuer to request a replacement check."

—Virginia Lemire

It had taken us five years to fulfill the objectives of the two-year Phase II $500,000 grant. We had "developed computer procedures for the early… remediation of dyslexia" and had "graphics, voice response, and touch-sensitive display." Using the Lexia software, a student received "verbal instructions while watching a video display and responding by touching various images on the screen."[15]

We were convinced that T&L would revolutionize the teaching of reading and that ASSESS would help teachers understand how their students were doing. Now we had to spread the word and sell them. There were so many students in so many schools who needed these products. Lexia had no money for a marketing campaign, so we had to invent a strategy. Looking back, we were not fully aware of how resistant the whole-language people were to a structured method of teaching reading like Lexia. Whole-language had a tight hold on reading education. Selling our products would prove to be much more difficult than we had expected.

---

15. Application Number 2 B44 BD19422-02, Computer Diagnosis and Remediation of Dyslexia, Summary Statement, page 2. 1985

"What was 'whole-language' teaching and why was it so entrenched?"

–BOB LEMIRE

Chapter 24
Whole Language and the Reading Wars

In the 1980s and 1990s, the "reading wars" were raging. Lexia was developing software that stressed phonics over whole language, and we believed that with our O-G method-based product we could change the way reading was taught. We understood that becoming a proficient reader required more than phonics, but we principally believed that without phonics there was little chance of a beginning reader progressing toward proficiency at an acceptable rate. Where we believed students needed explicit phonics instruction, whole language rejected phonics instruction altogether, believing that if students were just exposed to reading, they would learn to read. It was "phonics first" vs. "meaning first."

Even though the philosophy of whole language had been around for more than a century, it was fully entrenched as a teaching method when Lexia's phonics-based programs were coming to market.

I am not a "reading wars" historian. But I hope that everyone will come to understand exactly what the whole-language method of teaching means—and then perhaps understand the reason for many of the nation's reading problems. It will help explain why it was so difficult (and to some extent still is) to get Lexia products into the hands of students. To that end I have included excerpts from "The Great Debate Revisited," by Art Levine, published in the December 1994 *Atlantic Monthly*.[16]

In education no question has produced so much bitter debate for so long as this one: What is the best way to teach children to read? It is a critical issue, because there is clearly a need for drastic improvement in the way our schools do this essential job. As many as 20 percent of Americans above the age of sixteen are classified as functionally illiterate—unable to use print to perform essential tasks—and the ranks are growing every year. Even as the literacy crisis deepens, partisans of different reading approaches square off against each other. New teaching fads come and go.

The latest of these fads is known as the whole-language approach to literacy. It exposes children to interesting reading and writing at the expense of systematically teaching specific reading and writing skills. Whole-language teachers, for instance, encourage young students to recite along with them as the teachers read aloud from entertaining big-print books. One of their central beliefs is that language should be learned from "whole to part," with word-recognition skills being picked up by the child in the context of actual reading, writing, and "immersion" in a print-rich classroom. It is a philosophy that has won the backing of influential teachers' organizations, state and local education agencies, and tens of thousands of enthusiastic teachers.
...

The whole-language method is the latest incarnation of the "meaning-first" approach that has warred with the "phonics-first" approach to reading for more than a century. The traditional practice was to teach children the alphabetic code—the translation of abstract letters into sounds and words—before turning to actual reading. This approach was challenged in the mid-nineteenth century by Horace Mann, the secretary of the Massachusetts Board of Education. With a vehemence not unlike that of today's whole-language proponents, Mann denounced the letters of the alphabet as "bloodless, ghostly apparitions" that were responsible for "steeping [children's] faculties in lethargy." He argued that children should be taught whole, meaningful words first, and promot-

---

16. The entire article can be found at: http://www.theatlantic.com/past/politics/education/levine.htm

ed the "look-say" method that by the 1920s had come to dominate American education.

In the 1950s, however, the meaning-first approach was itself denounced, in bitter, still-resonant political terms, in Rudolf Flesch's best-selling book *Why Johnny Can't Read*. Flesch's broadside led to a wave of authoritative new studies. These concluded that reading programs that include systematic, intensive phonics instruction work better than those that do not. By the early 1970s most schools had returned to an essentially phonics-based program. But as the pendulum swung once again, these programs found themselves being criticized by some teachers and academics for killing off children's interest in reading.

Most of the critics initially directed their fire, with some justification, at conventional teaching tools. They denounced worksheets used to drill students with tedious rows of words, the "basal readers" with their dull Dick-and-Jane–style stories, and reading groups organized by ability which stigmatize children and often promote failure.

…

Whole language supposedly offers an easy way to literacy. Ken Goodman, a professor of education at the University of Arizona, is one of the whole-language movement's leading academics. He and his colleagues have argued over the years that learning written language can be as natural as acquiring spoken language, and that children can learn to read primarily by figuring out the meaning of words in context. "Good readers don't read word by word," Goodman says. "They construct meaning from the [entire] text." Indeed, "accuracy is not an essential goal of reading."

…

The whole-language approach is also favored by most teachers' colleges and university-level education programs. Its influence is so pervasive that in 1987 a survey of forty-three texts used to train teachers of reading found that none advocated systematic phonics instruction—and only nine even mentioned that there was a debate on the issue. Some form of the whole-language ideology has been adopted by more than a dozen state education agencies.

…

There have been troubling reports from the field about whole language. Reading scores of first-graders in San Diego dropped by about half when whole language was introduced there in 1990. One school-board member complained to the San Diego Union, "From parents I get lines like, 'You're experimenting with my kid.' I hope we didn't make a mistake—but I feel in my gut that we have."

Lexia believed that in the 1990s many teacher colleges no longer had faculty who knew how to prepare future teachers to teach phonics. But in 2008, when Lexia began offering lesson plans for teachers, we received enthusiastic support from teachers everywhere. Perhaps many were seeing explicit phonics lesson plans for the first time.

We at Lexia needed to learn about this "war" in order to demonstrate Lexia products to schools that used whole language, in addition to those that already used a phonics method. We soon learned that many whole-language schools thought of phonics as the "drill and kill" method and refused even to entertain the idea of Lexia's products. The very word "phonics" was not even a word we should use for our software as it was thought to have a significant negative impact on sales. Consequently, we had chosen the name "Touch and Learn Reading."

"One early conference was the International Reading Association Conference in Toronto, where Lit, Sharon, and I felt like aliens in a foreign land. It was a love-fest of whole language. Lexia's phonics-based reading programs were not welcome. Later, we regularly attended the Orton-Gillingham conferences, which were love-fests for Lexia's Touch and Learn and ASSESS."

–JUDITH COOLIDGE JONES

Whole-language teachers insisted that phonics drills suppress reading, but Lexia could prove otherwise. Nearly everyone who watched students using T&L said they loved it, and many teachers reported they had trouble taking students off Lexia because they had so much fun. They especially loved the novel repetitions that others said would be un-motivating.

I suspected that most elementary teachers had never experienced any reading difficulties of their own as children. And conversely, people who did have trouble learning to read generally did not become teachers. For the most part, teachers probably taught reading by the same method they learned to read. In the 1980s and 1990s, that was whole language.

The following satire is an amusing but real illustration of the effect of whole language on children. I found it on the Internet at The National Right to Read Foundation site.[17]

WHOLE LANGUAGE AT THE FORK IN THE ROAD

by Cathy Froggatt
Former NRRF North Carolina Director
Right to Read Report, February 1998

The purpose of this satire is to paint a clear picture of the anguish experienced by hundreds of thousands of young Americans as they advance through and leave school ill-equipped to handle the very real demands and requirements of school and life beyond. Cathy has heard many of these experiences first hand.

One day Dr. Goodguess died. The Gatekeeper to the afterlife told him that before entering the afterlife, he, like everyone, would be granted one wish to change one thing about his previous life on earth.

"What a wonderful surprise!" Dr. Goodguess exclaimed. "My greatest regret in life was that I didn't learn to read with Whole Language. As you undoubtedly know," he said, "I 'mainstreamed' that philosophy of reading into nearly every classroom in the English-speaking world."

"Your wish is granted," responded the Gatekeeper. "From this moment on, you will find that your brain has been altered. Now you will read the Whole Language way. You must now travel down the path you see before you for a short distance. There you will find a fork in

---

17. http://www.nrrf.org/satire_WL_at_Fork.html (Anyone may copy and distribute any information on this site as long as The National Right to Read Foundation at www.nrrf.org is given credit. The National Right to Read Foundation is a publicly supported 501 (c)(3) organization.)

the road. One path leads to Perdition, the other to Paradise. Signs are posted which clearly mark the paths. Choose carefully, because once you have chosen a path to travel, you can never turn back."

Dr. Goodguess marched off confidently until he reached the fork in the road. The left fork was marked with a sign that said: "Perdition." The road to the right said: "Paradise."

As he stood there, a look of puzzlement and then worry spread over his face. He scratched his head and thought, "They both start with 'P'; now what do I do? I've always been a risk-taker, but this is a frightfully important decision. I cannot make a mistake."

Just then, another founder of Whole Language, Dr. Sampler, died and stood before the Gatekeeper. "The hallmark of my life," he told the Gatekeeper with pride, "was the widespread influence my theories have had on reading instruction. I only wish that I had actually learned to read in a manner consistent with my theories: you know...naturally...without having to be forced to learn those low level phonics subskills."

His wish was immediately granted, and in a moment he joined Dr. Goodguess at the fork in the road. "Thank goodness you're here, Dr. Sampler," exclaimed Dr. Goodguess. "I am in dire need of some cooperative learning."

"Why, Dr. Goodguess, what is the matter? You look very distraught! What has happened to your self-esteem?"

"Well, Dr. Sampler, it's these darn words-in-isolation. You'd think there would be at least one picture clue somewhere?!"

"Hmmm, I see what you mean, Dr. Goodguess. Oh, no! Both signs have words that start with the same letter, and the words are about the same length."

As they stood pondering their dilemma, the earthly life of a College Professor of Education came to an end. As Professor Indoctrinate stood before the Gatekeeper, she stated with a rather high degree of confidence: "I have been completely happy with my earthly life. The life of a tenured professor, with the academic freedom it brings, was near perfect bliss. I wouldn't have changed a thing." ... .

When she arrived at the fork in the road Professor Indoctrinate, unwilling to provide any phonics information due to her thorough disdain for such "lower order subskills," encouraged Drs. Goodguess and Sampler to use the Whole Language cuing system they all knew so well. In an attempt to reassure them, she said, "Don't be upset if you can't read the signs just yet. After all, reading is developmental. In time it will all begin to click, maybe next year or the year after."... .

With the path behind them filling up with people impatiently awaiting their turn to pass through the fork, Drs. Goodguess and Sampler cried out in despair, "What we desperately need is more context!"

Just then they heard the soft spoken voice of a child. A little six-year-old boy walked up to them, looked at the sign on the right, and said with great pride, "I was taught to read with explicit, systematic phonics. I can sound out any word. The sign on the right says 'Paradise' and that's the way I'm going." And off he went.

Drs. Goodguess and Sampler looked at each other with knowing smirks. Their need for context had surely been met. Dr. Goodguess whispered excitedly, "Did you hear him say 'explicit, systematic phonics'? The path he took must be the road to Perdition! Quickly now, let's take the other path!"

As Lexia learned more about whole language, we sought out professionals who could provide us with whole-language/phonics comparisons. We began following Dr. Reid Lyon's assessment of the nation's reading problem and later invited him to speak at a Lexia conference.

The following biographical information is taken from Children of the Code, a social-education project that has compiled on the Web many interviews with people active in solving the reading problems in this country.

Dr. G. Reid Lyon is the former chief of the Child Development and Behavior Branch within the National Institute of Child Health and Human Development (NICHD) at the National Institutes of Health (NIH). He was responsible for the direction, development, and management of research programs in reading development, cognitive neuroscience, developmental psychology, behavioral pediatrics, language and attention disorders, human learning, and learning disorders.

Prior to his work at NIH, Dr. Lyon taught children with learning disabilities, serving as a classroom teacher and school psychologist for 12 years in public schools. He has authored, co-authored, and edited more than 100 journal articles, books, and book chapters addressing learning differences and disabilities in children. At the NICHD he was responsible for translating NIH scientific discoveries relevant to the health and education of children for the White House, the US Congress, and other governmental agencies. He also served as an advisor to President George W. Bush on issues related to child-development and education research and policies.

Early in Dr. Lyon's career, he taught third grade. In his own teacher-training courses he had learned that reading was a natural process and that you needed to provide kids with rich, interesting literature to motivate them, from which they would then infer the principles and move right along. He was taught not to deal with specific skills because they were de-motivating. Later, as a researcher, he found that phonics was essential to reading English. The sooner a student is taught to read using phonics, the better it is for the child.

The following are parts of an interview with Dr. Lyon available on the Web.[18] I think it extends the view of many regarding the teaching of reading in this country.

When we look at the kids who are having a tough time learning to read and we went through the statistics, thirty-eight percent nationally, disaggregate that, seventy percent of kids from poverty and so forth hit the wall. Ninety-five percent of those kids are instructional casualties. About five to six percent of those kids have what we call dyslexia or learning disabilities in reading. Ninety-five percent of the kids hitting the wall in learning to read are what we call NBT: Never Been Taught. They've probably been with teachers where the heart was in the right place, they've been with teachers who wanted the best for the kids, but they have been with teachers who cannot answer the questions: (i) What goes into reading, what does it take? (ii) Why do some kids have difficulty? (iii) How can we identify kids early and prevent it? (iv) How can we remediate [them]?

---

18. http://www.childrenofthecode.org/interviews/lyon.htm

Also in the interview, Dr. Lyon was asked why, with well documented science in place showing the success of the phonics method in teaching reading, whole-language was still being used in so many classrooms. He replied:

> The resistance in the educational community, particularly at the higher-education level where teachers are trained, is enormous, almost unbelievable. When you show people objective information, non-philosophically driven research that for these kids, these interactions work very productively such that where a youngster was at the tenth percentile in reading before, and is now at the sixtieth percentile in reading, and you can show that time after time, but you still see substantial resistance from the educational community, it begins to tell us that many of these issues are way beyond the kid issues, these are adult issues. They are fascinating adult issues where human beings are latching on to their beliefs, their assumptions, their egos and their careers rather than looking very clearly at what works, what doesn't, making sure people know what works, measuring it and getting the kids up to snuff.
>
> …
>
> My toughest challenge is not so much the science anymore. The toughest challenge we have is in moving the science to the development of teachers and their preparation, such that what they learn is actually objective and is based upon converging evidence rather than philosophies, belief systems, or appeals to authority. We need to get the information to teachers who have been, in a sense, propagandized into these very broad and general and non-evidentiary kinds of approaches that they use in teaching reading—absolute failures in terms of our scientific tests vis-à-vis their effectiveness.
>
> When I was a teacher I wanted to get up every morning and make a difference in kids' lives, and when I saw I wasn't making a difference in kids' lives that hurt the kids. But it also made me feel dumb, foolish, embarrassed. Our teachers sometimes feel the same way, but they can only teach what they've been taught, typically.

I read a story recently about the staggering casualties during the Civil War that occurred off the battlefield due principally to infections. The science of antiseptic methods was still years in the future but when science proved the benefits of antiseptic everyone adopted its use and deaths from infections were dramatically reduced. Today, science shows that explicit phonics instruction greatly decreases reading problems. One can only ask why, with science behind it, phonics instruction method has not yet completely replaced whole language.

Lexia could not directly challenge whole-language teachers and/or schools over their teaching methods. We could only promote our products as an aid to the student/teacher/school no matter what method of reading instruction was being used. Children would at least have a chance to become good readers if they were exposed to Lexia's reading software; better yet, students who had corresponding and supportive instruction would do even better. No matter what instructional method a teacher used, individualized practice was what mattered most for so many beginning readers. Where classroom teachers have very little time for one-on-one instruction and practice, many of their students who needed this type of attention suffered reading failure. Lexia's computerized reading products provided whatever amount of practice each student needed, without requiring constant teacher attention. How could anyone not want that advantage in every classroom?

"I was passionate about helping those kids. They deserved a shot at life, just like everyone else."

–Bo Lemire

Chapter 25
Bo Works with Teens

"After I graduated from college, in 1991, I spent the summer working in Ohio and Kentucky. I returned home in the fall and tried to figure out my next move. I assessed my skillset and decided to work with delinquent teens. I was experienced in this area, having been such a delinquent myself, and figured I had common ground with many of these individuals. My personal struggle with reading led to personal struggles in areas that would be helpful to me when working with at-risk and delinquent adolescents. In early 1992, I started at a company that worked with adjudicated youth from various juvenile courts in Pennsylvania. My first job was providing direct care to teenaged boys."

–Bo Lemire

Bo was quiet about the crimes of which these teens had been convicted. I once asked him for an example, and he told me that in one case a boy's father had shot and killed the mother before killing himself. Their 12-year-old son returned home to find his parents dead. He picked up the gun and ran out to rob a few stores. No wonder Bo was so engaged with these kids. They needed so much.

"I was willing to take risks for the kids I worked with; it showed them I was committed and not just doing a job."

—Bo Lemire

Each teen had been intercepted in the juvenile court and given the opportunity to attend "boot camp" rather than serve his sentence in the juvenile detention center. Bo was involved in their first experience at the camp. It was 24/7 supervision. I recall Bo saying that he placed his bed in front of the dorm door so no one could escape during the night. He remained in this position for 18 months.

"I was informed why these teens were sentenced and was sure that early educational failures played a role in their difficulties with the court system. Environment, mental health, and lack of responsible parenting were even more important causes."

—Bo Lemire

"Lexia had to become self-supporting and stop depending on venture capital."

—Bob Lemire

Chapter 26
Marketing Begins

Now that Lexia had something to sell, we had to find a way to sell it to whole-language schools as well as those that were not. We designed T&L to be a supplement to any curriculum a teacher used, and we had named it carefully so that whole-language people would not reject it for its title.

Judith Coolidge Jones continued to display Lexia products at language conferences. In 1990, she had booths at seven conferences from New Hampshire to Washington, DC. We also placed our first ad in a special-education issue of *Media & Methods*.

Many schools had no computers at all, and if they did they had little budget for software. But Lexia had been founded with grant money—so we had the idea that we might help schools write grants to acquire their own hardware and software. We spent a good deal of time assisting schools in this way, and although this did help with a few sales, it was not going to get us where we needed to be.

We needed to broaden the sales and marketing effort and do it soon. Lexia had to stand on its own. In late 1991, Lexia hired BJ Clemson to lead sales. Soon thereafter, Judith Coolidge Jones and Nancy Johnson hosted a Lexia booth in the lobby of an Orton Society conference and returned with dozens of product-interest cards, indicating that the signers wanted more information about Lexia products. We were thrilled with this response, even though telephone responses were going to take time. BJ called her friend Sue Newbury.

> "BJ offered me a table, chair, pencil, phone, and no pay for at least five months. I had a couple of shoe boxes for storage, as there were no file cabinets."
>
> —SUE NEWBURY

Sue knew little about computers, dyslexia, or the Lexia products ASSESS and Touch and Learn. BJ provided her with a framed one-page product description. That was it for training.

> "My first call was to someone who denied making any request for information. The second call was to a school's new tech person. This person asked questions to which she should have known the answers and shouldn't have been asking me, so I asked her to make a deal. She would tell me what questions I should ask about the computers, etc. at her school and I would tell her what kind of questions she should ask a sales rep. This was my first sales training!"
>
> —SUE NEWBURY

In many schools at that time the only person who purchased software was the tech person, no matter what the subject or what teachers might want. If Sue called the school and asked for the tech person, she might or might not get connected because secretaries were told not to put through sales calls. Somehow Sue got a copy of the Quality Education Data (QED) book for New England, which listed all the schools and the full names of the administrators and teachers. She could now call the school for the target person by name. Lexia gave Sue business cards that said "Sue Read," hoping it would be easier for potential customers to understand and remember.

The early versions of T&L were not easy to install and support on unknown computers. School computers generally ran more than one program, and other programs (or some hardware components) often interfered with Lexia. Programmer Nancy Johnson had to be available to support installations and field ongoing questions from teachers.

> "I had finished my graduate degree and worked at Raytheon Data Systems until they closed. Bob asked if I could help at Lexia for a few months before looking for another job. I hesitantly agreed and stayed a couple of years. Our two children were away at school so I could focus on Lexia. There was no money to pay me, so I joined Bob as an unpaid employee."
>
> —Virginia Lemire

BJ felt very strongly that if a school or teacher just tried Lexia software, they would buy it. So Sue's next goal was to get someone at a school to say they would take T&L on a trial basis, and Lexia would ship out the entire product. We sent about 100 boxes and waited. Sue followed up on each trial, either to get payment or to get a return. Most were returned. Months later we sent bill collectors to two or three recipients. This was an expensive way to get a product into a school. And it was unsuccessful largely because there was no one to help the recipient install it and no one to explain or demonstrate the broad functionality of the product. Written information could only go so far, and technology-averse teachers were reluctant to spend the time and energy needed to explore the functionality completely. We would learn later how very important a salesperson and trainer were for a successful sale.

Not only was this lesson an expensive one for Lexia, but it also opened us to the possibility that our software could be stolen and/or copied. There was not much we could do about a school buying one copy and replicating it; we just had to believe that most of our customers were honest. We did set up a database of our paying customers and the software they had purchased.

Our concern was heightened after receiving a call from a computer-repair shop. They had a personal computer from a tutor who said it had had Lexia loaded but needed it reloaded after a disk crash. We checked Lexia's records for the computer's owner and found that the tutor had been shipped a trial and had returned it. We received similar calls from schools or classrooms where there were no records of their having purchased it.

Everyone in the software industry was struggling with theft/copying issues. Our products were still being shipped on floppy disks for MS-DOS computers, and there were few options for protection. We put information in our shipping boxes about illegal duplication and encouraged users to abide by the law, but that was really the only thing we could do; more aggressive measures would make our products much harder to use.

Besides doing direct sales, Sue Newbury started to do presentations in area schools. It wasn't easy. A demo required her to lug all the equipment from the office, handling it gingerly lest it fail to operate during the demo. If she needed support, Nancy was always in the office—if Sue could find a phone. Oftentimes Sue would bring food to the demos, just to get the teachers to attend. She even had to exert a little scolding to get an audience of teachers to be quiet and focus on her presentation.

"I started work at Lexia packing boxes. It was not long before Lexia needed someone to present the products on a day when Sue Newbury was already scheduled, so they sent me. I had a terrible fear of public speaking so had written out what I wanted to say. I arrived at the school an hour early. As it happened, there was a cemetery across the street, so I walked in circles around the gravestones rehearsing my talk. It didn't help. I was still terrified. Thirty teachers showed up to hear me. Over time, I learned how to speak. Now, as a Lexia reseller, I can talk about it for hours."

AARON SOMOZA

Sue then had the idea that the prison population likely contained many nonreaders, and she made an appointment at a local prison.

> "Going through security, I was required to show my business card and driver's license and sharply told the guards not to touch my stuff! The guard was not happy when he saw different names on the card and license. I explained about 'Sue Read' and waited. Finally one guard laughed and said the story was so unreal that it had to be OK. I was admitted."
>
> —SUE NEWBURY

Sue reported that the prison demo was very significant for her. One prisoner asked to have an individual hands-on demo, so the guard reluctantly stood back to let him try. The prisoner learned for the first time that the sound of an a was like in apple. She said a tear ran down his face when he thought he might eventually learn to read.

Sue went on to give teacher training in area schools, train new reps at the home office, and represent Lexia at reading conferences. Despite her good efforts, her relative impact on overall sales was limited: we needed 20 people like Sue, but did not have the money.

How were we going to penetrate the school market?

The "wagon train" on its trek

"Bo was so isolated from us that it felt like we were on different planets."

Chapter 27
Bo Supervises Teens

It was not long before Bo was assigned to the "wagon train." This moving group of teens, their wranglers, and staff including cooks moved up the eastern states during the summer and returned south for the winter. The "train" consisted of 10 actual covered wagons, each pulled by two mules, with outriders on horseback, a semi-trailer truck with supplies, and a cook shack, as well as a trailer with portable toilets. Two teenage boys drove each wagon while others rode horseback. If a boy had misbehaved the day before, he walked. Each night the train camped at a pre-arranged and pre-set-up location. Before long, Bo was the assistant director.

"I was transferred to the company's wagon train, which started north from Florida in the early spring. All the teens had been through 'basic training' and had done well enough to be promoted to the wagon train. The environment was completely different from the streets in Pittsburgh or Philadelphia. We used dependency on the animals and the group as a whole to teach responsibility and cooperation. This was intense 24x7 work because I lived where I worked."

—BO LEMIRE

We heard from Bo only when he was near a phone, which on the wagon train was not often. We could call an 800 number to find out where he was and relay a message for him, but contact was limited.

"Our wagon train was observed passing through South Boston, VA, one day and someone asked us if we wanted to take our wagons out on the NASCAR race track. We opened the race with mule-drawn wagons. The fans at the race got something special that day, and the boys enjoyed it."

–Bo Lemire

The boys who could accept their place on the wagon train did well. Those who constantly challenged authority or tried to manipulate their environment did not do as well. They had spent their lives manipulating someone or something, and changing that behavior was very hard to do.

"It was not unusual for boys to attempt running away from the wagon train. I ran after more than a few, sometimes across backyards or across farm fields. Runaways were sent back to the company's permanent facilities. I was good at this hands-on care, but it was not easy to believe that they could turn their lives around."

–Bo Lemire

"I realized that Lexia had a product with a completely new approach to reading."

—WAYNE McKNELLY

Chapter 28
Resellers

In the early 1990s, Wayne McKnelly saw one of Lexia's small ads and called me from his office in Texas. He and his business colleague, Bruce McComas, were selling educational products in the school marketplace for a Houston company. They had come across Lexia and realized that we had a completely new approach to reading—and they wanted to sell it. We helped Wayne and Bruce start a few Lexia trials in each of their territories.

Sometime later, I visited their employer in Houston, who showed little interest in Lexia. Like so many others, he didn't seem to understand how Lexia's reading software worked because he knew little about the acquisition of reading skills. His work was strictly with catalogs, and his selling method did not include making calls on school customers. It wasn't long before Wayne and Bruce called me, saying that they were doing all the area selling and asked if they could have their own territories. I gave them the go-ahead.

One of our first installations outside of New England was a large sale by Wayne McKnelly to the Edinburg, TX school district. Wayne asked Lexia to provide a trainer on site. We didn't have one, but by this time Virginia knew enough about the product that she could pinch-hit. We flew to Texas and she gave a half-day training session to thirty teachers, who responded positively to a software teaching method that was completely new to them.

Wayne took us to another Lexia school in San Benito, TX where we saw thirty or more computers set up in a lab. It was quite thrilling to see Lexia in use on all those computers.

Bruce McComas's first sale was to the Benito Martinez Elementary School in El Paso, TX. The school had a very good principal who was open to new ideas. Each student used Lexia in a lab setting one day a week. The teachers liked T&L very much and the students loved it, but installing it had not been easy. Full installation required about fourteen disks per computer, and Bruce reported there was almost always one or more disks that didn't work. Lexia immediately sent replacements upon request—but again we learned that a salesperson or trainer was essential for the installation and continued use of the software.

Sales were painfully slow. No matter how good the product was, it could not sell itself. We desperately needed a new strategy. Perhaps a consultant could be hired to come to Lexia, review our products, and tell us how to sell them. No one remembers how we found Rosie Bogo, and when we recently spoke with her about this book she didn't remember either. She had founded Hartley Courseware, which published educational software, and had recently sold her 150-employee company to Jostens. We arranged for her to visit Lexia.

Rosie Bogo visited us for two days and recommended the use of resellers to reach more schools and increase sales. These independent individuals could sell our products without being on the Lexia payroll; they would work on commission. Could this be our new strategy? It sounded good to me. Before leaving Lexia, Rosie gave me her list of preferred resellers, located all across the US. They all knew Rosie and knew their own territories, schools, and administrators. She let us use her name to contact them.

"I spent a day or two at Lexia in the early 1990s. I recall seeing Nancy, their only full-time programmer, packing boxes and thought it was a poor use of a talented resource. But then, we had done the same thing in the early days of Hartley."

—ROSIE BOGO SWART

With this list of resellers, we again thought our ship had come in. We would think this over and over again as each opportunity came to us, and eventually learned that every step and every event helped shape Lexia's future. We knew we were on the leading edge of educational software, reading products, and especially phonics-based reading, and nothing was going to stop us.

We set about contacting Rosie's resellers and hoped they would want to market our products. We sent the Lexia products to them and supported them while they studied them. We followed up with all the resellers we had contacted and visited several.

One trip took us to Florida. We met Virginia Stoner and Ken Hodges of Educational Learning Systems (ELS), in Tallahassee, FL. They represented various companies focusing on reading. Virginia and Ken had set up their computer in a Tallahassee hotel. We went through the program with them, explained how it branched and created reports.

> "We liked Bob Lemire's dedication and the product's functionality and were favorably impressed with Lexia's mission-driven philosophy to help every child learn to read. We liked the simplicity of the screens but were concerned that the depth and breadth of the functionality were not easily seen."
>
> —VIRGINIA STONER AND KEN HODGES

Looking back, perhaps Lexia should have spent more time training the resellers about the full functionalities behind the simple screens. Some told us much later that they were very slow to learn the full functionality of Lexia products.

Both Virginia and Ken knew that their territory was not yet ready for a phonics-based product. Whole language was well entrenched. They felt that Lexia was a pioneer in reading software whose need and acceptance were still in the future. Nevertheless, they got good results in the schools where they were able to sell it as a special-needs product. Gradually schools and state educators began to realize that children were failing reading and that instruction needed to be changed. This gave ELS a greater opportunity to sell Lexia products.

When Bruce McComas became a Lexia reseller, he knew nothing about teaching reading. In fact, he had had his own problems with reading.

"I did not read well but was good at math. Early in my college years I visited the college bookstore to see what academic major required the least reading. I decided on physics and taught it for a few years after graduation. We all want to study and choose careers in areas we find fun and easy to learn. Why would someone want to teach elementary school, especially the lower grades, if they had trouble learning to read? If a teacher had learned to read by 'just getting it,' she/he may expect her/his students to do the same."

–BRUCE MCCOMAS

As an authorized reseller of Lexia products, Bruce began specializing in reading. He was describing the three reading publishers he represented to an El Paso teacher when she asked him which one was the best. He replied that he didn't know, he just sold them, and she told him that if he was selling reading products, he should know more about reading. He asked her to recommend books on reading that would help him—and he reported that as soon as he began to understand the complexity of reading acquisition, he was hooked. This teacher became his reading mentor. When Lexia's Strategies for Older Students (SOS) became available, Bruce installed it on his mentor's computer and asked her to give him feedback. She thought it was terrific and wanted it in her school. The research Bruce had done on reading acquisition, as well as his mentor's enthusiasm for the Lexia method, gave him great confidence in showing Lexia products to other potential customers. As a Lexia reseller, he had a passion for Lexia that was based on his own struggle—and he didn't want others to go through the pain that he had experienced.

We signed up resellers for most territories in the US. Their ability to sell our products varied with their territory's acceptance of structured reading methods. A man from England contacted us who has since become a star reseller; he sold so much that we eventually made a slightly different version for students who speak, read, and write with a British flavor—or, as they would put it, flavour.

The Lexia software demo CD

"They would sell our product if they could put their name on it."

–Bob Lemire

Chapter 29
Phonics Based Reading

To increase Lexia's exposure to every elementary school in America, we explored listing our products in educational catalogs. I knew of a nearby catalog company that served the school trade, so I contacted them in 1994 and introduced myself and Lexia's products. I learned that they were interested both in entering the educational-software business and impressed with what they had heard about Lexia. They invited me to speak to their upcoming sales conference.

"I distinctly remember Bob Lemire's presentation at the sales conference. His passion was completely on display: I could see how utterly committed he was to the vision of using software to help children learn to read. Almost no one in the audience understood that vision. The sales force was accustomed to selling workbooks from a catalog. They did not appreciate or understand software and were generally dismissive. It would be a long road for Bob Lemire."

–Paul Schwarz

A few days after my presentation, the company's president called and said he would like to include T&L in his catalog but that it had to have their name on it and had to be upgraded to run on Microsoft Windows and Macintosh. At that time, Lexia products ran only on MS-DOS. It was true that schools were just starting to purchase Macs and PCs that ran Windows, and we knew we had to keep pace with new technology, but there was no time available from the Lexia developers and no money to outsource such a large project. The catalog company was willing to commit its money for the upgrade if Lexia would agree to a contract that would license Lexia's intellectual property to them. They could then put their own name on it and sell it as their product. Although cautious, we were excited about the possibility of getting a Windows/Mac version on CD-ROM without putting out any of our very limited money. Was this the way to get it done? Who would do all the work?

The catalog company suggested that Lexia contact a young Boston company that did Windows contract work for the business community. We made an appointment with this company and showed them the DOS-based Touch and Learn Levels I, II, and III, along with all their documentation. They showed us some of the software they had created under contract for other companies. We were amazed by the dazzling artwork they said they created with a software tool called Director. We were very eager to have a new version of T&L that would include this excellent animation and art, and their engineers were very interested in doing work for Lexia (although they were not much interested in the user manual and documentation). They spent little time running the software before submitting a bid for $78,000.

Lexia struck a deal with the catalog company in 1995 to repay the $78,000 through a royalty schedule. The contract also specified that Lexia could cancel the agreement at any time and pay back any remaining debt.

Lexia handed the MS-DOS–based code to the contractor along with all of its support materials. Fortunately for us, Virginia was available to supervise the contract: she had done formal software development at Raytheon and knew the process well. The young software engineers doing the contract work had little use for a structured development process. They had no method or strategy for reporting and fixing bugs, nor did they have a software-control system. Version control was nearly nonexistent, and new versions came to

Lexia two or more times a week, mostly hand-delivered as there were no electronic means at the time.

Virginia implemented a bug-reporting system and generally did the in-house quality checking. She quickly found that the developers had little idea of how wide and deep the product's functionality was and how much branching it contained. They had all the documentation but were not using it. The number of bug reports multiplied with every software delivery. Weeks and months went by, and we began to think the project would never be completed. We desperately needed this new release, as we had stopped any enhancements or bug fixes to our MS-DOS version.

> "The contractor had seriously underestimated the amount of work required to replicate the code on Windows/Mac. Their management became threatening during one of my visits and I fled their offices. Finally, the project manager left the company and took our project with him. I'm pleased to say that he completed the contract."
>
> —VIRGINIA LEMIRE

The new Windows/Mac version of Lexia's reading software looked and felt entirely different from Touch and Learn. The graphics and color were greatly improved. We loved it, and the students who used it during beta trials loved it too. This all-new-looking version needed a new name. We were seeing the word "phonics" more often in the educational world and decided to call it Phonics Based Reading—PBR.

Not long after the catalog company launched its version of Lexia's software, their president sent me a memo with the results of one of their salesperson's introduction of the software to a school. The teachers had opined that it seemed a very weak program. Their complaint was that the word choices, while displayed on the screen, were not read out loud. They used the example of a picture of a wet hen and three corresponding phrases, only one of which said "a wet hen"; they were disturbed that the program did not read all three phrases aloud to the student. The salesperson felt this product would be hard to sell.

We could not believe it! Both the salesperson and the teachers were looking at our product as if it were a whole-language program and all the students had to do to learn to read was be read to. This was a complete misunderstanding of our product. How would the student learn to read if the program did the reading? Could these teachers understand that this was a program to practice reading, not practice listening to someone else read? Would this vast misperception be a common problem with our products? What other misunderstandings would people have? Perhaps this was the beginning of our realization that Lexia products would be best sold by fully trained representatives—salespersons who were educated in reading acquisition and in how the Lexia product was particularly designed to teach and practice reading.

With the final and successful delivery of PBR, Lexia took over control of the computer code. It was now our job to maintain the code for bug fixes and provide enhancements. The Windows version also presented the opportunity for copy protection. There were several encryption copy-protection schemes available that could be built into the software, but we soon learned that encryption technology would not work with Director, the software tool used for our animation and art. Director was a product of Macromedia, so I studied their website for hints on how to proceed. I soon learned that one of their board members was yet another man with whom I had ridden the train when he lived in Lincoln years before. I wrote him a letter and we soon had a software fix for Director that allowed Lexia to add encryption copy protection. We breathed a little easier about unauthorized propagation of our products.

With the new computer code now at Lexia, our developers created a demonstration CD-ROM of Lexia's PBR program. The CD was shipped to hundreds of potential customers who had shown interest in our reading program either by telephone or by filling out a card at a meeting or show. We created a large data base of interested tutors, teachers, and schools.

The language consultants moved on to design Lexia reading products for preschoolers and older students. The preschool product focused on phonemic awareness, learning the alphabet, and rhyming. Older students received advanced levels of vocabulary, graphics, and stories.

Everyone at Lexia was thrilled with Phonics Based Reading. The interface to the student was so much better. The student and teacher reports enhanced the product greatly. Product installation and stability problems were solved with the CD-ROM version of PBR. Lexia now had its products running on Windows and Mac computers!

We were sure this was the moment we had been waiting for. Surely sales growth would follow.

Bo became Program Director aboard the tall ship *Bill of Rights*

"There could be no doubt as to who was in charge."

–Bo Lemire

Chapter 30
Bo on the *Bill of Rights*

While Bo was working on the wagon train, he began looking for his next promotion within the company. Soon he was made program director of the tall ship *Bill of Rights*, the company's 136-foot two-masted schooner.

"I boarded the ship for the first time when the teens were below deck watching a movie on the only television set. Twice the boys were called to the main deck to meet me, but they ignored the request. I went below, picked up the TV, carried it on deck, and threw it overboard. That was their introduction to their new program director.

"When top leadership positions change in youth corrections, clients (the teens in my case) hope for a power shift. It was very important to immediately show dominance, both physically and mentally. There could be no doubt as to who was in charge. Respect was earned by actions.

"In a detention center or jail, on a ship or in a classroom, authority, respect, and discipline are required for the group to function for everyone's benefit. These boys had committed crimes and were sentenced accordingly. We were attempting to teach cooperation, responsibility, and self-discipline. On the streets they were used to being judged by the colors they wore, or the wad of cash they carried, or the gun in their pocket. Respect and power came from dominating others. We rewarded the boys as they shifted their value system based on the safe environment we provided. I shared common ground with these kids. I understood what "less than" felt like. I could offer a way out of that mindset by helping them take care of a tall ship."

—Bo Lemire

Bo had a good deal of responsibility. The ship was equipped for sisteen boys, several childcare staff, and a cook. It was commanded by a captain and a first mate. Bo was in charge of the overall program, with a teacher as his educational assistant. The teens had regular school classes as well as seamanship training. Each week the ship made for a port where a FedEx delivery was waiting with a check. After cashing the check, Bo could take a select number of boys to buy food and do laundry.

"It was always fun to see people's reactions when we arrived at the check-out with ten or more shopping carts of food."

—Bo Lemire

Virginia and I visited Bo on the *Bill of Rights* when it tied up for a few days at the Chesapeake Bay Maritime Museum. It was a very impressive ship. We saw Bo's two-bunk stateroom that he shared with his dog. The dog slept on the lower bunk and would not let anyone in the cabin except Bo.

"We visited ports all up and down the eastern seaboard, including Savannah, Charleston, Norfolk, and Atlantic City. We sailed by the Statue of Liberty, visited the fishermen in New Bedford, MA. We spent time in Gloucester, MA watching the greased-pole contest. We sometimes had blue whales follow us. We got to see and experience things most people dream about. We created memories these kids will take with them and never forget. I was very proud of these sailors as they finished their sentences and left the ship. I was more than a little concerned that they were going right back to the place they came from. Would their experience with us be enough to change their lives?"

-BO LEMIRE

The only difficult time Bo reported at sea was during a storm when one of his adult crew fell out of his bunk and got a deep six-inch gash on his head. After administering first aid, they changed course and headed to the nearest port some seven hours away.

Bo was twenty-seven years old.

The schooner *Bill of Rights*   copyright and courtesy of Susan Hunt Williams

Reseller conferences: [top] at Seiji Ozawa Hall at the Boston Symphony Orchestra's Tanglewood Music Center in Stockbridge, MA; and [bottom] the Aldrich Mansion on Narragansett Bay in Warwick, RI.

"Resellers did not like seeing Lexia's product with another company's name on it."

—Bob Lemire

Chapter 31
Reseller Conferences

Lexia needed a way to bring its resellers together to share information about our products, provide product training, and provide and teach the use of our marketing materials. One of the best reasons for getting resellers together, however, was to introduce them to one another and build a collective team spirit.

Lexia hosted its first reseller conference in the late 1990s, with seven resellers attending. Over the years the conference grew in size and was hosted in Boston as well as other selected places in New England. Resellers arrived from all parts of the US, Canada, England, and New Zealand. They loved visiting Boston and loved the conference. Most attended annually.

Every year the conference included one or more presentations from teachers, principals, or superintendents who told us in detail how they implemented Lexia's reading software and monitored its use, and how their student reading scores had improved. It didn't matter if it was a school district where English was a second language, a city with multiple home languages, or a homogeneous suburb with English as a first and only language: Lexia had a very positive effect on students' reading scores. These shared experiences helped us understand how and why the Lexia software was purchased, how it was deployed, and how the salesperson continued to support each sale. All presenters spoke of their students' love for Lexia.

It was inspirational to hear the gratitude of administrators not only for the effectiveness of the reading product and increased scores among their students, but also for the ease of getting students to use it.

> "I sat next to John Dicker at his first Lexia reseller conference. We chatted about the history of Lexia. His company, Greenfield Learning, soon became and continues to be one of the top Lexia resellers in the country."
>
> —VIRGINIA LEMIRE

On the final evening of each conference, Lexia provided a dinner event at a special location like Fenway Park, the Boston Public Library, the John F. Kennedy Library, the Boston Symphony, or on a cruise of Boston Harbor. Many Lexia employees were able to attend conference sessions and special events, which gave resellers the opportunity to meet the people with whom they had spoken on the telephone. It was very expensive to sponsor these reseller conferences, but we felt they were essential for our marketing and team-building efforts.

> "I got my job at Lexia after my wife saw a small help-wanted notice at Donelan's Market in Lincoln. My first day coincided with the annual reseller conference. I knew next to nothing about the job I had just taken and little about Lexia. I walked into the conference room during a presentation. Everyone was so attentive, dedicated, and enthusiastic. They sounded so devoted to the products and what they were doing. I had never been in a meeting like that. I wondered if I had been hired by a cult!"
>
> —BRIAN ROFF

It was only a matter of time, of course, before the catalog company selling their version of our software collided with our Lexia resellers. The latter complained that they were running into the Lexia product in their territories with another company's name on it, and the catalog company's version was actually competing with Lexia's Phonics Based Reading. The only thing to do was to terminate the catalog contract. Lexia ended the contract in 1999 subject to its terms and paid the catalog company about $125,000 over a three-year period. One of the results of this action was that the resellers had a great deal more respect for Lexia Learning Systems because we had responded to their concern.

At Lexia, sales were increasing thanks to the resellers. They were very excited about Phonics Based Reading—they had something new to sell and received a good deal of support from us to help in their efforts. We were beginning to feel more secure about the future of Lexia.

Mott Meeks and his younger brother with father Dr. Littleton "Lit" Meeks

"School was pure hell!"

– MOTT MEEKS

Chapter 32
Another Dyslexic Boy

What must it have been like for a dyslexic boy who was never properly diagnosed and never received the appropriate help he needed to become a successful reader? Virginia and I learned about this boy in 2012 when we interviewed Lit Meeks for this book. Lit's son, Mott, was visiting his father that day and told us about his life with a reading disability.

Mott Meeks was born in 1947 and received his early elementary education in the Atlanta, GA school system. He told us that school was pure hell. He recalled that his teacher scheduled a child to read aloud each day. When it was Mott's turn, he would attempt to hide his poor reading skills by being absent or by wetting his pants so he would have to go to the lavatory. He felt stupid and alone. His parents took him to see one psychiatrist after another in an attempt to find out what was wrong. Nothing was ever found.

In 1958, Lit moved his family to Lexington, MA where Mott entered fifth grade. Lit and his wife informed the school of Mott's reading problems and he began receiving extra help. Mott described the in-school help as more of the same instruction he was getting in the classroom. At the same time his parents arranged for additional help in the form of a private after-school tutor. Mott told us that the tutor used a phonetic approach but little practice time was available in just three hour-long after-school sessions per week. The private tutoring and in-school help ended after eighth grade with little to show for it.

Like the Lincoln schools, Lexington was unable to understand reading disabilities and how to teach these children to read. Students with significant reading disabilities, like Mott, were vastly underserved.

In each new class or school he received grades of F, but later, when a teacher realized that something was wrong, he would be given a C/F grade. Failing in front of his friends was devastating. No matter how many things he did successfully, it did not overcome his feeling of inadequacy.

As he grew up, he excelled in functional spatial relationships. He entered the Boston Architectural Center to study architecture. He left school in his thesis year and began teaching art in a school for disabled children. The school's headmaster told him that he was dyslexic. That was when he finally knew the cause of his reading failure.

Ultimately he used his spatial skills to find a very responsible job in the construction business; reading and spelling are not essential. Mott married and, with his wife, Catherine, raised a family. Although he is very successful, he does wonder what choices he might have had or what he might have become if he had not been dyslexic—or had been remediated.

Bo was born eleven years after Mott. We lived in Lincoln where, on the train platform, I happened to find the right doctor to diagnose and remediate our son.

"We simply got off mission."

–Bob Lemire

Chapter 33
Remaining True to the Mission

I stepped down from running daily operations at Lexia in 1997 but stayed on the Board of Directors, serving as its chairman. Lexia expanded its management team.

When I was asked how I could pass the leadership to new people when I had founded the company that was "my baby," I replied that I had learned this in my early 30s when I captained the Boston rugby team I had helped found. A new player had arrived from England who showed superb playing and leadership skills. I recruited him to replace me as captain and we soon began to win more games. That experience had a profound effect on me. Many companies go astray when their founder is unable to step aside and hire others to run the show. I am grateful that I learned this early in my life.

The first member of the new team was Jonathan Bower, hired as CEO. He arrived with his own PC and we provided him with a desk and a telephone.

> "When Bob Lemire showed me the Lexia reading products, my impression was that Lexia was 'a diamond in the rough.' I knew how much the products were needed and thought that Lexia was years ahead of its time. They were not fancy products but they focused on the lessons without distracting bells and whistles."
>
> –Jon Bower

In January 2001, Paul More became the next member of the team as Lexia's financial controller and head of not only the finance department but also human resources, facilities and grounds, and everything else that was not sales or development.

As the number of Lexia employees, contractors, and projects increased, Lexia occupied more and more space on Lewis Street in Lincoln. These included the three-story mansard-roofed former residence at #2; where, when a train went by, any empty chair rolled to the center of the room; the former garage at #11A, with its very cold floors; the brown building at #11; the white house at #9; the former cobbler shop across the parking lot from #2; a few rooms in the former pickle factory/town barn at #20; and the former hair salon at the side of #2. All these locations had different phone numbers and none were connected by a common network. Paul obtained permission from Verizon to hang wires on their poles, pulling twisted pairs for the phones and fiber-optic cable for a network to all the buildings, where they were threaded through basements, crawl spaces, and knee walls. Finally, all locations were connected. These offices were our startup space.

Later in 2001, Lexia hired Joel Brown as the next new member of the management team, to be VP of sales. Joel had never worked with resellers but he quickly took up the challenge. He invited seven Lexia resellers to serve on a Reseller Advisory Board that met three or four times a year, to act as a sounding board for new Lexia products, sales, and training ideas. They became important advisors to Lexia.

"When I was new at Lexia, Bob Lemire told me that if a teacher in a tiny school on some Maine island called to say she had a couple of kids who needed Lexia, Lexia would pay one of its employees to make a visit to the school (or ask the reseller to go). Even though Lexia would never make any money with that customer, it was the right thing to do for children who needed help. Bob said that was the Lexia mission.

"He also told me that the biggest problem with an under-achieving salesperson was not the loss of sales but all the children who were being denied the use of Lexia."

—JOHN KONVALINKA

In 2001, after passing Lexia's street sign for nine years, Lincoln resident Bob McCabe walked through our door and asked what the company did. With his background in education, he soon became a Lexia employee assigned to conduct research on student usage of Lexia software that we believed would help us learn more about how students performed on our products. We could learn how students responded to each activity and unit and how difficult or easy each exercise was, all of which would help us when changes were planned or activities redesigned.

Bob McCabe's first major project was a company-developed research project in the Revere, MA public schools, where they agreed to allow a study, with a control group, of the efficacy of Lexia products in their elementary schools. Revere was a working-class community near Boston where perhaps a third of the students lived in non–English-speaking homes. It was a perfect school district in which to do a study. In the course of the research, Revere allowed Lexia into five school buildings, where hundreds of students were pre-tested and post-tested. Lexia's PBR was put into thirty classrooms where many teachers had never used technology and needed training. There were both control classes, where the students did not use Lexia, and experimental classes, where Lexia was used. At the end of the year, the Revere data revealed that the already-identified at-risk Title I[19] students who used Lexia were very much ahead of the students who had not used Lexia. This study was the basis for Lexia's first peer-reviewed publication. The outcome of this research was our affirmation that the Lexia mission could be fulfilled.

> "I was in the Revere classrooms often, where my six-foot ten-inch frame was easily recognized by teachers and students alike. One day a first- or second-grade boy called me over to where he was using Lexia's Phonic Based Reading. He moved the headphones from one ear and pointed at the screen, saying, "This looks like a game but this is work!"
>
> —BOB MCCABE

---

19. Enacted in 1965, Title I of the Elementary and Secondary Education Act (ESEA) provides the resources to schools to enable students to reach proficiency as determined by the assessment of state standards in reading and math. Such schools are situated in low income communities which struggle to provide a high quality education to all children. *National Title I Association*

The Revere research took more than four years from its initial design, in 2001, to its publication in 2006, and was only the beginning of Lexia's extensive research. It set us on the path of continued, committed scientific evaluation of our reading programs. It made Lexia one of the most rigorously researched, independently evaluated, and respected English-reading programs available.

Over time, Lexia's research team published six peer-reviewed papers on the efficacy of Lexia products: in 2006, the *Journal of Research in Reading*; in 2007, *Perspectives on Language and Literacy*; in 2008 and 2011, *Reading Psychology*; in 2009, the *European Journal of Special Needs Education*; and in 2011, the *Bilingual Research Journal*.

Lexia's research studies have all been peer-reviewed. This means that the work has been independently evaluated by one or more professionals in a relevant field, often as a prerequisite for publication in a particular journal. Any company can say it has research that shows positive results. But the customer needs to see what type of research it is. If it has not been published in respected professional journals that require peer review, it can mean little.

Through research, Lexia's PBR was found to accelerate the development of critical foundational literacy skills in the early grades. Lexia's Strategies for Older Students has also been proven effective in remediating struggling readers in middle and high schools. The research studies followed rigorous scientific standards including the use of control groups, pre-testing and post-testing, standardized and norm-referenced reading tests, and stringent statistical data analysis.

Researched confirmations of the gains students made using Lexia's reading products were important to have for business reasons. The unsolicited endorsements that arrived by mail or email were also important to us: they were written by parents, adult learners, teachers, tutors, and young users, and they were sometimes heart-wrenching. The following user stories are but three of the many we received.

"A woman called and asked about Lexia's reading product for her 80-year-old father, who had never learned to read. He had asked his daughter for help because his wife wasn't supportive. The daughter bought Lexia's Reading's Strategies for Older Students for him. Many months later, she called back and told me that her father had worked on Lexia at her house for many hours nearly every day, often taking naps when he got tired and then going right back to it. Sometime later, her father and mother went out for dinner. When the menu arrived, her mother started to read it to her husband, as usual. He said that he didn't need her to do that, and he proceeded to read the menu and order dinner on his own. Her mother was astounded and, of course, she and her father couldn't have been happier."

—SUSAN KANO

Dear teacher,

Thank you for introducing us to the Lexia program. Without it, my husband and I would never have fully understood the areas where Joe was struggling or the exact kind of help that he receives from you in class (like all of those sound cues he likes to teach me). We love that the school was able to offer us the Lexia program at home because we were able to see first hand how quickly his reading improved every time he used the program. It seems that as he went on more and more each night, he started to fly through the levels and gain all the skills he was missing and with that, the self confidence and self assurance that he can indeed read! We love you and the wonderful things you have done for Joe and we love the Lexia program and all it has done to keep up and improve Joe's reading skills!"

—JOE'S MOM AND DAD

> "I used Lexia to teach the lowest achievers and was able to get most of them on grade level! When using Lexia up close in a small group, I came to realize the program was tracking the students' problems that I was noting on paper during the Lexia session, down to the actual letter symbol causing them the most confusion. I've become a believer! These low-achieving students already felt like failures. Your program not only gives them hope, but it allows them the drill and repetition needed to re-train their brain wiring, hold the skill, and generalize it! You use external visual and auditory rewards but you also create internal rewards simultaneously— genius, absolutely genius!"
>
> –MRS. MCKECHNIE, LESLIE FOX KEYSER ELEMENTARY SCHOOL

In the late 1990s and early 2000s, Lexia embarked on a number of projects that were not all within our mission. They were cash-hungry projects, and they nearly brought Lexia down with financial problems before they were canceled or sold.

The largest one was Cross-Trainer, which computerized a tutorial to improve cognitive skills and promised excellent results for students. We enthusiastically embarked on this project, even though it was completely outside of our mission. It was developed under a $2 million grant[20] but was still incomplete when the grant funding ran out. After pouring a substantial amount of money into it, we discovered that it would be very hard to sell because schools did not want to take responsibility for their students' cognitive skills. Although Cross-Trainer was moderately successful, key members of the development team began to move on, and Lexia's ability to update and support it was crippled. Eventually Cross-Trainer was discontinued.

Lexia summer camp was a smaller project created to test the emerging Cross-Trainer product and help struggling readers. Students and parents loved the camp but it was not contributing to Lexia's cash flow and when money got tight, it was canceled.

---

20. The grant was secured from the Advanced Technology Program of the National Institute of Standards and Technology of the US Commerce Department.

The Partnership for Achievement in Reading (PAR) was another large project we started under the Massachusetts "Reading First" federal initiative. In 2003, Lexia won a four-year contract from the Massachusetts Department of Education for $2.5 million per year to train teachers in scientifically based teaching methods. Although the state seemed very satisfied with our work, after a year they decided to take the contract back into their Department, leaving Lexia with $171,000 of unreimbursed expenses. This was a big hit.

At the request of one of Lexia's resellers, we started another project, the Comprehensive Reading Test (CRT). It could be used in only a few states and was canceled when it developed database problems that required additional money to fix.

At the same time, Lexia began work to "Lexia-ize" two paper products. One was Creative Reading Instruction[21] and the other was the Herman Method, an Orton-Gillingham compliant paper-teaching system. Neither of these products was expected to bring in much money, so when money got tight, Creative Reading Instruction was canceled and the Herman Method was sold.

When projects were canceled and the number of employees was reduced, everyone worried about the company's future. Looking back, I could see that we had simply lost focus on our mission. We were sure that we had the key to teaching reading in the English-speaking world; we thought that only Lexia could change this world and that we had to be all things to all people. A larger company might have been able to do all this, but not our emerging company with its limited available funds.

> "The importance of the mission was what gave us the strength to endure the stresses and strains of building Lexia into a viable force in the market."
>
> —PAUL MORE

---

21. Authored by Sharon Weiss-Kapp

It was some time before expenses were reduced enough and sales were robust enough that Lexia could overcome its loss of focus and related financial problems. Lexia recommitted to its mission of teaching every beginning and struggling English-language student to read.

By 2006, Paul More could no longer be asked to worry about lightning hitting any of the several Lewis Street buildings containing our irreplaceable computer systems, the snow removal, and the cleaning service. Lexia needed modern facilities, even though some of our thirty employees feared that leaving the startup space might adversely affect our corporate culture. Paul found cost-effective space in an office park in nearby Concord, MA that preserved the natural cross-lighting and individualized layout that had characterized our environment in Lincoln. Everyone came to enjoy clean restrooms and not having to worry about whether there would be electric power that day. Lexia was back in the business of developing and marketing K–12 foundational reading software.

Lexia was then in good financial condition with excellent facilities. The product line was shown to be effective. The sales channel was well developed. The company was growing and the future was bright. It is always important, however, to remember how you got where you are and what you want to become. Lexia's mission should always be the first consideration in all decisions.

So now you have it. The early history of Lexia Learning Systems encompassed many years of struggle and uncertainty in our quest to end reading failure in the English-speaking world. There is, of course, still much work to be done.

My task is over but my heart will always be with Lexia.

"This story started with our son Bo being diagnosed with dyslexia. It seems fitting that he finish the story."

—Bob Lemire

Chapter 34
Completing the Circle

In Bo's words—

In 1997, I left the *Bill of Rights* and briefly returned to Lincoln, where I packed everything I owned and headed west. It was time for a new challenge and I decided I would look for it in the West. I had no job lined up and no particular destination, but I was a risk-taker and it sounded like a good idea. I drove west on Interstate 70 and when I saw the Rocky Mountains I stopped and looked for a place to live.

I worked several jobs in residential treatment centers for adolescents, spending the last few years in this field as an adolescent caseworker for one of Colorado's Department of Human Services organizations.

> "When Virginia and I visited Bo in Denver, he took us for a drive into a neighborhood where several of the young boys he worked with lived. He saw that they were out after curfew probably because they had so little supervision at home. He expressed the frustration of ever being able to improve their lives."
>
> —Bob Lemire

As I grew older, got married, and started a family of my own, I began to rethink my career path. I was good at my job, but the success rate for making a lasting impact on these kids' lives was low. Success was measured in inches, and failure was common. It is much easier to prevent major lifelong problems than to fix them after they have existed for years. I decided I wanted to prevent kids from becoming delinquent. I knew that many of the individuals I worked with for fifteen years had trouble early on in school. I decided to change my focus and stop trying to repair the broken lives of teenagers and focus on improving the lives of young students so that they wouldn't have to be fixed later.

I began to think of Lexia. I had spent years putting together my current career, and making a change would be a significant event. I struggled with the decision of getting involved with Lexia because everyone would know I was the son of the founder. I wanted others to value me for my experience and contributions, not my last name.

In 2004, I decided it finally made sense for me to look to Lexia for a new career. I called the VP of Sales at Lexia to see if there might be a way to get involved. I was put in touch with the reseller who covered Colorado. I had no experience with sales but knew a lot about reading problems, based on personal experience. I was a good communicator and influencer based on my experience working with adjudicated youth. It was not long before I was offered a job representing Lexia in the state of Colorado. The sales skills needed to succeed took time to develop, and I am grateful I was given ample time to adapt to this new job.

Providing help to all young readers became my mission. I try to share my experiences as a struggling student and use them to help teachers and administrators with their students. I try to explain why explicit instruction works better than handing a struggling student more books to read. I try to help schools and districts think about how best to implement and utilize the Lexia programs in a meaningful way. I try to help teachers and administrators make an impact on their student learning that they may have never known was possible. I try to tell the struggling students whom I meet that there is a way to become a good reader.

What about the on-grade-level students? How much better can these students get if they are taught better, earlier, faster? How much of their future knowledge is delivered by the written word? Everyone can get better at this. Lexia is not a program just for dyslexic students, or just for students needing intervention; it is something that all elementary students can use to become successful readers. Advanced students can advance faster. Let's set the bar high for all students.

Lexia is, quite simply, a tool that can create life-long opportunities for students. Parents won't have to see their kids struggle with reading, won't have to watch those struggles snowball into other problems. The children will just grow up reading, simple as that. Grade-level students will advance further and move faster than previously thought possible.

In November 2012, when my father resigned from Lexia's Board of Directors, I asked to join that group to represent not only the family's minority interests but resellers as well. Lexia's management and board members welcomed me into that role. I had completed the circle from being the reason the company was started to participating in shaping its future.

There is no greater job satisfaction than positively impacting the lives of children. It is impossible to measure how many students may have had serious lifelong problems had it not been for Lexia. Success, to me, is measured by the impact we make on students' lives. We must never forget that there will always be another student who needs Lexia.

To do this work, it has to make sense—you have to believe it—you have to feel it in your heart.

Lexia makes a big splash at Fenway Park at the 2012 sales conference

Acknowledgments

Although I am a founder of Lexia Learning Systems, there are many people who helped make Lexia a reality. Without Dr. Edwin Cole and Dr. Littleton Meeks there would be no Lexia. I don't believe Lexia could have been started without Dr. Cole's deep knowledge and experience with the O-G method of teaching reading. He generously gave us his advice and encouragement. In 1994, I visited him on his 90th birthday. We talked about how he had helped hundreds of his patients learn to read and that now the software that he had helped create would teach millions of students to read.

Lit Meeks brought the enquiring mind of a scientist to implement my dream of ending reading failure in our schools. Alice Garside brought her day-to-day practical knowledge and use of O-G to the design of Lexia's products. I regret that neither she, Dr. Cole, nor Lit Meeks has lived to see Lexia's enormous success.

There were few other places in the US that had the expertise that the Boston area had for the acquisition of reading and the remediation of reading disabilities. Lexia had help from the following O-G language consultants: Alice Garside, Pamela Hook, Sandra Jones, Sharon Marsh, Helen Popp, Barbara Porter, Faith Rugo, Sharon Weiss-Kapp, Isabelle Wesley, and Angela Wilkins. Each of these consultants made significant contributions to the design of Lexia reading software that would provide the scaffolding needed to help students learn to read. They shared their passion for helping readers at all levels and generously gave their experience, ideas, and time to make sure that Lexia's products contained the best possible pedagogy. We at Lexia are forever grateful for their contributions. Other professional guides included Mary Chatillon and Elizabeth E. White. Many thanks go to past and present Lexia employees who brought their skills to the mission of teaching all beginning and struggling students how to read.

There are many people who helped make this book a reality. My spouse Virginia and I interviewed many current and former Lexia employees, including Perry Bent, Jonathan Bower, Joel Brown, Jeff Dieffenbach, Tony Ferro, Nick Gaehde, Debbie Gillespie, Mary Ann Hales, Pamela Hook, Nancy Johnson, Judith Coolidge Jones, Susan Kano, John Konvalinka, Sharon Marsh, Catherine Meeks, Lit Meeks, Paul More, Sue Newbury, Kathy O'Loughlin, Beth Pilgrim, Elliott Reinert, Brian Roff, Paul Schwarz, Aaron Somoza, and David Stevens. They all were informative and helpful. Thank you for sharing with us your Lexia memories.

We also interviewed early resellers including Wayne McKnelly, Virginia Stoner, Ken Hodges, and Bruce McComas. Their memories of the early days of Lexia have been invaluable. Thank you for sharing your recollections.

We interviewed Mary Ellen Whitaker and Steve Olenick of Audio Link, where the Lexia sound was and is still recorded. We spoke to Earl Oremus, an educator and Lexia Board of Directors member; early sales consultant Rosie Bogo Swart; Dr. Cole's daughter Abbey Cole Dawkins; and language consultant Sally Grimes. Mott Meeks consented to talk about his life as a dyslexic person. All of these people helped us greatly.

Thanks to Nick Gaehde for asking me to write about the mission and history of Lexia. Thanks also to Lexia's Marc Gemma, project lead and Michael Bergen for the book cover design.

I am most appreciative of the editing help from Mimsy Beckwith, along with Mary Ann Hales, Kathy Gosselin, and Elise Lemire. Thanks especially to Virginia for sharing the weight of my commitment to write this book.

Most of all I am grateful to our son, Bo, for being willing to talk and write about his experience of growing up with a reading disability. We learned so much from him and are so proud that he has transformed a former disability into his own mission to change the way reading is taught in the English-speaking world.

Lexia's Various Lewis Street Offices Throughout the Years

Cobbler's Shop

Old Town Barn

Number 11 Lewis St.

Number 2 Lewis St.

Index

List of Illustrations